HOOFED MAMMALS
OF THE WORLD

HOOFED MAMMAL

PUBLISHED BY CHARLES SCRIBNER'S SONS NEW YORK

OF THE WORLD

BY UGO MOCHI AND T. DONALD CARTER

*In memory of
the great life work of
my brother*

DOCTOR ALBERTO MOCHI

ACKNOWLEDGMENTS

My FIRST impulse and desire was to dedicate my studies in this book to my very dear friends Edgar and Cecile Hearty whose encouragement and understanding of my art contributed to the completion of this collection. Truthfully the meaning of my first impulse still remains.

However, the death of my brother Alberto in 1949, and the revelation of his great life work, toward the building of a cure for our troubled society, inspired me to remember him here.

I owe more than I can say to all those whose main interest is the study and understanding of animal life—the many zoological gardens and natural history museums, both here and abroad. But in particular, I am grateful to staff members of the American Museum of Natural History and especially to Hazel Gay, head librarian, for her most efficient and courteous help.

To Don Carter I would express my gratitude for his assistance and splendid cooperation and for the privilege of having been associated with him in this work. I am ever thankful that this adventure of the "Hoofed Mammals of the World" has been the means of my fully appreciating a man whose modesty and altruism belie at first the extent of his vast knowledge.

Admiration is due my dear wife Edna for her patience, understanding and encouragement during my years of activity on this work.

May I also mention here three men who in my early years inspired and guided me in the creative field of Animals in Art: the painter, Professor Paul Mayerheim, and the sculptor August Gaul, under whose supervision I studied and worked two years; also, Dr. L. Heck, Director of Berlin Zoologischen Garten.

My first and final indebtedness goes to my mother, who during my childhood in Firenze, Italy, so keenly taught me to love and appreciate the beauty of nature and the great masterpieces of art.

UGO MOCHI

FOR THEIR interest and help, I would like to express my thanks to the following: to my sister, Miriam Carter Conn, for reading the text and offering much valuable aid; to my wife, Helen E. Carter, for her patience and help in checking the manuscript; to Dr. Harold E. Anthony and Hobart M. Van Deusen of the Department of Mammals of The American Museum of Natural History for their instructive advice and cooperation, and to Martha I. Cotter, who not only typed the text but also gave me helpful suggestions.

T. DONALD CARTER

CONTENTS

FOREWORD TO THE NEW EDITION

Hoofed Mammals of the World was first published in 1953. Ostensibly a systematic collection of animal silhouettes by Ugo Mochi and supporting species accounts by T. Donald Carter, its extraordinary combination of art and natural history immediately made it a collector's item. The original edition measured an unwieldy 12 by 15 inches. This new one is more manageable, although it retains the integrity of the original illustrations' scale of 1/32 life size. The book is still unique in its zoological coverage and in its animal portrayals.

Man began scratching the outlines of other animals on bone, painting them on cave walls, and forming their shapes in wood and clay as soon as he became man. Nevertheless, silhouette, called the "art of outline" by Mochi, at first seems unsuitable to convey the character of vigorously three-dimensional creatures like zebras and giraffes, deer and rhinoceroses. Such is the artist's skill that his "shadows" assume movement and perspective before our eyes.

Mochi's tools are simply a lithographer's knife and black paper. Nothing could be flatter and no medium could more promptly reveal a flaw in the artist's interpretation of posture or proportion. That flaws are undetectable is a tribute to naturalist Carter as well as artist Mochi, for the work was a collaborative effort from the start.

I find that Mochi's animals often elicit that special thrill of sudden insight known to those who enjoy distinguishing one creature from another. They bring into focus diagnostic characters missed in viewing the living animal itself. How this subtle effect is achieved is unclear, but I suspect that the very perfection of the Mochi outline combined with the unlikely medium of the "shadow" art holds the secret. The "shadow" stimulates the viewer's eye in a way foreign to conventional animal portraits, perhaps in the same fashion that a book is so often better than its cinematic reproductions.

Ugo Mochi was born in Florence, Italy, in 1889. Childhood interests in animals and art developed into professional studies at the Art Academy in Berlin, where he spent hours modeling animals at the zoo. By 1928, when he settled in the United States, Mochi had relinquished not only his sculpture but also a promising singing career to concentrate upon the "art of outline." His work, covering an enormous variety of subjects, has appeared in countless articles, in architectural decor and in many books. But it is in animal illustrations that his art reaches its highest development: medium and subject so blended by long concentration that Mochi's animal "outlines" are not likely ever to be equalled.

T. Donald Carter was born in New Jersey in 1893. By the time he was twenty he was making trips to Central and South America, seeking animals for the New York Zoological Society. Before he was twenty-five he joined the staff of the mammalogy department of the American Museum of Natural History where he eventually became assistant curator. He has studied mammals and birds on expeditions from Florida to Alaska, in Brazil, Venezuela, and British Guiana; in Ethiopia, Rhodesia, Cameroun, and former French Equatorial Africa; as well as in China, Indochina, and Labrador. To this book, he brings extraordinary experience, unsurpassed love of his subject, and clear informative prose.

Aside from its esthetic appeal, *Hoofed Mammals of the World* is unusual in its zoological coverage. It remains the only review of the world's hoofed animals within one volume. This seems strange, for civilization, as we know it, could not have evolved were it not for man's beasts of burden and his herds and flocks of cattle and sheep, all derived from the wild progenitors, or their relatives, portrayed and described by Mochi and Carter.

In the final analysis, a book like *Hoofed Mammals of the World* is more than a work of art or a reference; it is a tribute to nature and a testimony of admiration from two men of unique talents and experience.

Mochi and Carter have depicted and described more than 290 hoofed mammals—all of the known species and many of the races. Today, more than ninety are listed as rare or endangered. For some, it will be only a decade or two before Mochi's "shadows" have more substance than their subjects.

WILLIAM G. CONWAY
Director,
New York Zoological Society

PREFACE

WHEN THIS cooperative work had been accepted by the publisher, Ugo Mochi wrote me, "This book, for my part, is the culmination of a life-long interest in the beauty of animals."

It is unfortunate that every reader cannot see the originals of Mr. Mochi's illustrations, for not only was each one made with a knife instead of a pen but furthermore each illustration was cut out of a single piece of paper. It is most remarkable that anyone could possess the delicacy of touch and the infinite patience to do such work. But when the results are absolutely correct in every detail of pattern and outline and are exactly to scale, the work approaches the miraculous.

A misnomer, natural but misleading, has frequently been applied to the type of art used in this book. As the reader will note in even a casual glance at the illustrations, the animals appear to be alive. Unlike the silhouette, a name sometimes erroneously used to describe Ugo Mochi's work, his knife lifts the subject from the flatness and simplicity of this shadow portrait, so popular in the middle of the eighteenth century, to the realm of creative art. His subjects, instead of being mere outline, are seen in relation to space, and give the impression of life and movement. The true form is the main character distinguishing an animal, color and shade being secondary. Ugo Mochi's work emphasizes the form.

Where others use brush and paint to give color and form, Ugo Mochi with a knife alone, suggests shade and motion. His genius is well exemplified in this book by the way in which, first, through outline form alone he reveals the subtle differences between closely related animals; and second, through the addition of other animals placed in the distance, he creates the effect of space. In such groups, the largest animal is portrayed on the scale on one inch to thirty-two inches which is also used in all pictures of individual animals.

I first met Ugo Mochi in the spring of 1950, when he appeared at the American Museum of Natural History and asked if he might inspect some giraffe skins in the Museum's collection. His knowledge of mammals was instantly apparent. Upon completing his careful studies he explained that the beauty of animals had always been one of his chief interests and that he had spent many years studying their form and reproducing it in sculpture as well as in this, his favorite form of art.

At this time he was planning a children's book about the capture of a giraffe, to be illustrated by a series of his famous cut-outs. He later brought in the series and some of his illustrations of antelopes, a few of which he thought might be used in conjunction with the story. The exactness of the form, the skill and beauty of the work so impressed me that I immediately suggested that these extra illustrations should not be used in a children's book, but should be placed where adults could appreciate their quality. We thereupon decided that he should complete this series of all the more important antelopes. Later we enlarged the plan to include all the hoofed mammals except the domesticated ones. Upon the completion of his enormous task Ugo Mochi honored me by asking if I would write an explanatory text to accompany his plates.

The hoofed mammals or ungulates, as they are known among the scientists, are without doubt the most important group of animals in relation to man. Man has been closely associated with them throughout his entire existence. A number of forms were readily domesticated and for centuries these have been bred for food, clothing, transportation, and many other purposes. Man is dependent on the ungulates for most of the world's meat supply, milk and its by-products, wool for clothing, leather, skins for fur coats, bristles for brushes, glue and fertilizer. Before the advent of the machine age members of this group were used almost exclusively as draft, pack and riding animals, and even now there are places where they are indispensable.

The hoofed mammals also afford recreation to many thousands, for the majority of big game trophies come from this group.

But surprisingly, the public seems to know very little about this important group as a whole and, aside from scientific works, no attempt appears to have been made to bring them all together in one volume.

There are about 940 recognized forms of modern ungulates, including the numerous subspecies. To include them all would not be practical in a book of this scope. Most species are here represented and those not illustrated are almost identical in form and structure with a close relative which is pictured. The slight difference is chiefly in variation in color or some other minute detail.

Although to many the Latinized names may seem superfluous for a book of this type, they are really necessary. Over many parts of an animal's range a number of English or common names may be used. A steinbock in Europe is an ibex but in Africa a steinbok is one of the small antelopes. Our American puma or mountain lion is found from British Columbia to southern South America and throughout this great stretch of country it is known by over forty common names. A gopher in parts of this country is a burrowing rodent with large and formidable incisor teeth, but in many parts of the West and Midwest it is a ground squirrel, while in Florida it is a tortoise. Scientific names are much more standardized and a particular name refers to a specific animal the world over, no matter what language is used in any particular country.

Without the use of scientific names the classification of animals, which constitutes their relationship to one another, would be difficult. As it is our hope that this book will be of interest to numerous people not especially interested in the scientific study of these animals, we have tried to keep the technical terms down to a minimum. However, a brief description of the main divisions and their characteristics should be of interest.

The ungulates have been divided into two main groups, the artiodactyls or even-toed ungulates, and the perissodactyls or odd-toed ungulates. The former are further characterized by having the first digit wanting and the third and fourth digits the longest, with the terminal phalanges flattened and encased in a hoof. Their grinding teeth have broad crowns with ridged surfaces to aid in the grinding of their herbivorous diet. The artiodactyls include the giraffes, deer, pronghorn antelope, antelopes, goat-antelopes, musk oxen, takins, goats, sheep, oxen, hippopotamuses, pigs, peccaries, camels and chevrotains. All the above with the exception of the hippopotamuses, pigs, peccaries, camels and chevrotains are placed in a group called the Pecora or cud-chewers. This group and the chevrotains lack the upper incisor teeth. The perissodactyls have the center digit the longest but the number of functional digits may vary. The terminal phalanges of these toes are flattened and encased in a hoof. Like those of the artiodactyls, the molars are broad with ridged surfaces but differ in minor details. All have the upper incisor teeth. All ungulates lack the clavicle or collar bone. The perissodactyls include the horses, zebras, asses, rhinoceroses, and tapirs.

The artiodactyls are again divided into nine families: Giraffidae, the giraffes and okapi, characterized by having the short horns covered with hair-bearing skin; Cervidae, the deer, animals carrying solid antlers made of bone, which are shed and renewed annually, though there are two exceptions where the male carries canine tusks instead of antlers; Antilocapridae, the pronghorn antelope, in which the horns carry a prong, grow over a bony core and are renewed each year; Bovidae, the antelopes, goat-antelopes, musk oxen, takins, goats, sheep and oxen, in which the horns grow over a bony core and are never shed; Hippopotamidae, the hippopotamuses, with a large rounded muzzle and the canine teeth greatly enlarged to form tusks; Suidae, the pigs, with a mobile snout terminating in an oval disc and with canine teeth greatly enlarged to form tusks, the upper ones curving upwards; Tagassuidae, the peccaries, with a snout similar to the pig's but with the upper tusks growing in a downward direction and one toe, instead of the usual two, on the rear of each hind foot; Camelidae, the camels, animals which walk on broad fleshy pads and have two toes on each foot, which terminate in claw-like hoofs; Tragulidae, the chevrotains, small animals similar to the Pecora in not having upper incisors and in chewing their cud but different in that the stomach has only three chambers in place of four, and, in the case of the males, having long tusk-like canines.

The perissodactyls are divided into three families: Equidae, the horses, zebras and asses, which in the modern forms have but one toe touching the ground; Rhinocerotidae, the rhinoceroses with three well-developed toes on each foot, the foot resting on a fleshy pad, and with an upright horn or horns on the nose; Tapiridae, the tapirs, with elongated nose, rather rounded facial profile, with four toes of each fore foot and but three toes of each hind foot touching the ground, and like the rhinoceroses, with the feet resting on a fleshy pad.

The families are again divided into genera which are divided into species. Species are subdivided into subspecies. Roughly speaking a member of a genus differs from other members of the family in one or sometimes more, important characteristics while the species under a genus shows less important differences. A subspecies is a division in which environment has caused a slight change, such as in color or size, from the typical species. In the scientific name of a mammal the first name always denotes the genus and is always capitalized, the second name is the species, and the third, when used, is the subspecies. The two latter names always begin with a small letter.

The principal sources used in the preparation of this book are listed in the bibliography.

T. DONALD CARTER

HOOFED MAMMALS
OF THE WORLD

1
THE GIRAFFES
AND THE OKAPI

NUBIAN GIRAFFE (*Giraffa camelopardalis camelopardalis*)

Of all the animals to be found on the African plains, the giraffe is the most impressive. Whether it be a herd hurrying off at their peculiar swinging gallop or a solitary male looking down from his towering height (large male giraffes have been recorded over eighteen feet in height) one feels that at last he has observed the ultimate in nature's display of mammals. While many animals are dependent on their noses and ears to locate their enemies, the giraffe trusts to its sight, and very little escapes the notice of this walking observation tower. Its height also saves it from the keen competition of other browsing animals. The food of the giraffe consists almost exclusively of the leaves and small twigs of the acacia trees, the flat-topped thorn trees scattered so commonly over the African plains. Other animals feed from the lower branches but the topmost leaves are reserved for the giraffe.

The giraffe prefers the plains and the thornbush country and shuns the forests and swamps and a river forms an impassable barrier. Although the giraffe ordinarily drinks regularly, it is capable of going without water for long periods. It generally goes about in small herds but occasionally a solitary individual will be found. A single calf is the rule though twins have been reported.

Most scientists agree that there are twelve subspecies of the giraffe occurring within its extensive range, from the Sudan on the north to the Orange River in South Africa on the south.

Beginning at the north we have the Nubian giraffe, the first giraffe to be scientifically described. This animal is found in Nubia, the eastern Sudan and Ethiopia. The chestnut-colored spots are large and distinctly quadrangular in shape, separated by buffy-white lines. The lower legs are relatively free from spots. Farther west and south in the Sudan at Kordofan occurs the Kordofan giraffe. The chief difference between this giraffe and the Nubian giraffe is the small size of the spots on the upper legs. Farther south and east in Kenya from Mount Kenya and the Tana River northward into Ethiopia is found a very different type of giraffe, the reticulated giraffe—considered a full species by many. This giraffe has very large rufous blotches separated by very narrow white lines or reticulations. In northwestern Kenya and southeastern Uganda is one of the largest of the giraffes, the Baringo giraffe. Similar in many ways to the reticulated giraffe, the white lines of the markings are wider and the neck may be blotched instead of narrowly reticulated. As with the Nubian race, the legs lack any markings below the knees and hocks. In northern Uganda the South Lado giraffe is found. This giraffe is intermediate in most characteristics between the Nigerian and the Baringo giraffes. Throughout most of Kenya and Tanzania occurs the Masai giraffe. The spotting of this animal is jagged and irregular in outline. The lower portion of the legs shows some degree of spotting. To the west in Nigeria and sparingly west to Senegal a very large giraffe occurs, the Nigerian giraffe. It is paler in color than the Nubian giraffe and with more numerous spots. A characteristic feature is a large fawn-colored patch below the ear. In the northeastern Congo, the Congo giraffe is found. This animal has the well-spotted legs which are characteristic of the more southern giraffes, and the large spots of the body are subquadrangular. Like the northern forms, this subspecies carries a

4

well-defined horn on its forehead. The Angola giraffe has similar markings but with ill-defined margins, and differs in having but a low swelling on the forehead instead of the well-developed anterior horn. The Transvaal giraffe is restricted to the northern Transvaal and is the form found in the Kruger National Park. Its body color is dark chocolate-brown and the body spots are broken up into irregular star-like spots. The Rhodesian giraffe resembles the Cape giraffe but the ground color is darker and the anterior horn is well-developed. The Cape giraffe has spots more or less quadrangular in shape and widely separated by lighter ground color. This type of coloration is known as the blotched type and is the very reverse of the netted or reticulatd type of the north. The legs are fully spotted to the hoofs. At the present time the giraffe does not exist south of the Orange River in South Africa.

Another characteristic which distinguishes the northern giraffes from the southern forms is a large bony protuberance on the forehead, in some subspecies forming a third horn. This is especially well-marked in the Baringo giraffe, and old males of this form often have two more posterior horns, small horns immediately behind the main pair, arising from the occipital bones, thus giving the giraffe a set of five horns. In the southern subspecies, the Angola, the Transvaal and the Cape giraffes, the frontal horn is rudimentary and shows but a slight bump on the forehead. The Rhodesian giraffe, although one of the southern forms, has a pronounced anterior horn.

The color patterns of the different forms of giraffes may be divided into five different types of markings. (1) The spotted type includes the Nubian giraffe, the Kordofan giraffe, the Nigerian giraffe, the Angola giraffe and the Congo giraffe. (2) The reticulated type includes the reticulated giraffe and the Baringo giraffe. (3) The jagged type includes the Masai giraffe, becoming more definitely of a leafy type in the Rhodesian giraffe. The similarity of the Cape and the Rhodesian giraffes is not so much in the type of the spots as in their coordination. (4) The star-like type is characteristic of the Transvaal giraffe. (5) The blotched type includes the Cape giraffe. This blotching is caused by the intermingling of the lighter edges of the spots with the darker, uneven background. These five types are not always constant and giraffes from the same locality show some variance in their markings. On this account, not all scientists agree on the twelve subspecies. Some

OKAPI (*Okapia johnstoni*)

reduce the number of subspecies to eight, making *cottoni* a synonym of *rothschildi, congoensis* a synonym *of antiquorum,* and *infumata* a synonym of *angolensis,* and regarding *wardi* and *capensis* as synonymous.

The only living relative of the giraffe is the okapi, a forest animal living in the great rain forest of the Congo. Unlike the giraffe, only the males bear the short hair-covered horns.

The okapi is about five feet at the shoulder and is of a dark brown color with white stripes on its flanks and legs. The existence of some strange forest animal in the eastern Congo was known as far back as 1890, but it was not until 1901 that Sir Harry Johnston sent portions of the striped skin to the British Museum. On account of the markings it was first thought that a forest zebra had been discovered but soon its true status was determined.

NUBIAN GIRAFFE *(Giraffa camelopardalis camelopardalis)*

KORDOFAN GIRAFFE *(Giraffa camelopardalis antiquorur*

9

RETICULATED GIRAFFE *(Giraffa camelopardalis reticulata)*

RINGO GIRAFFE *(Giraffa camelopardalis rothschildi)*

MOCHI

NIGERIAN GIRAFFE *(Giraffa camelopardalis peralta)*   SOUTH LADO GIRAFF

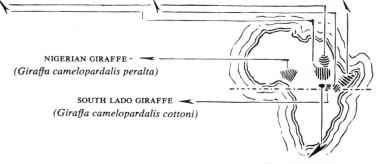

RDOFAN GIRAFFE
(...affa camelopardalis antiquorum)

BARINGO GIRAFFE
(Giraffa camelopardalis rothschildi)

RETICULATED GIRAFFE
(Giraffa camelopardalis reticulata)

NIGERIAN GIRAFFE -
(Giraffa camelopardalis peralta)

SOUTH LADO GIRAFFE
(Giraffa camelopardalis cottoni)

CONGO GIRAFFE
(Giraffa camelopardalis congoensis)

11

...a camelopardalis cottoni)   **CONGO GIRAFFE** *(Giraffa camelopardalis congoensis)*

12

MASAI GIRAFFE *(Giraffa camelopardalis tippelskirchi)*

AN GIRAFFE *(Giraffa camelopardalis infumata)*   TRANSVAAL GIRAFFE *(Giraffa camelopardalis wardi)*

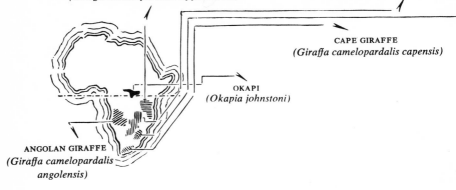

MASAI GIRAFFE
(Giraffa camelopardalis tippelskirchi)

RHODESIAN GIRAFFE
(Giraffa camelopardalis infumata)

CAPE GIRAFFE
(Giraffa camelopardalis capensis)

OKAPI
(Okapia johnstoni)

ANGOLAN GIRAFFE
(Giraffa camelopardalis angolensis)

14

ANGOLAN GIRAFFE *(Giraffa camelopardalis angolensis)*

TRANSVAAL GIRAFFE
(Giraffa camelopardalis wardi)

OKAPI *(Okapia johnstoni)*   CAPE GIRAFFE *(Giraffa camelopardalis capensis)*

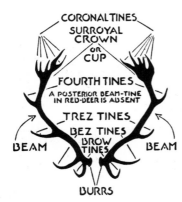

RED-DEER ANTLERS
EIGHTEEN POINTS

CORONAL TINES
SURROYAL
CROWN
or
CUP

FOURTH TINES
A POSTERIOR BEAM-TINE
IN RED-DEER IS ABSENT

TREZ TINES
BEZ TINES
BROW
TINES
BEAM

BEAM

BURRS

REMARKABLE SET OF RED DEER ANTLERS

2
THE RED DEER
AND THEIR ALLIES

The remarkable set of red deer antlers illustrated is from the outstanding collection preserved in the King of Saxony's Castle at Moritzburg, near Dresden, Germany. This collection contains antlers of the red deer of the fifteenth, sixteenth, and seventeenth centuries, and shows the remarkable heads obtained at that time.

The red deer is chiefly a forest- or mountain-inhabiting animal and generally goes about in small herds. The antler usually has at least five tines with the brow tine forming an obtuse angle with the beam; tail short, general color uniform, generally with a lighter rump. The young are spotted. This is one of the best-known of the deer and is found throughout most of Europe, Asia Minor, northern Iran and North Africa. A number of subspecies have been named, chiefly on account of size, color and horn formation. The typical red deer is the one found in Sweden and is one of the larger forms.

The number of points varies on different antlers but there are generally five or more on a side. When a stag has twelve points in all he is known as a "Royal," and if there are fourteen he is known as a "Wilson." At the top of the main beam the points frequently form a cup.

The red deer has been introduced into New Zealand.

SCOTTISH RED DEER *(Cervus elaphus scoticus)*

SCOTTISH RED DEER. Slightly smaller and darker than the typical red deer of Sweden. It is lighter in color in the summer with a distinct border to the lighter rump patch. General color dark reddish brown, grayer on the face and neck. Legs blackish brown. Long hair on neck in the winter coat. Brow and bez tines generally close together and at a distance above the burr. Found today in the moorlands of Scotland, in Westmoreland, Devon and Somerset. A few still occur in the New Forest and in Kerry. Most of the park deer of England are derived from the larger subspecies imported from the mainland.

WESTERN EUROPEAN RED DEER. The red deer of western Europe is a large deer equal to the typical red deer of Sweden. It is a light-colored deer, differing from the typical red deer in having the light-colored rump patch bordered with black. The finest horns come from the heads of this red deer. Found from France, Germany and Denmark to the western

18

WESTERN EUROPEAN RED DEER *(Cervus elaphus hippelaphus)*

CORSICAN RED DEER *(Cervus elaphus corsicanus)*

Carpathians. A smaller, grayer subspecies of the red deer, *C. e. hispanicus,* is found in Spain; while on the west coast of Norway *C. e. atlanticus* occurs. This deer is smaller in size and paler in color than the neighboring typical deer.

CORSICAN RED DEER. One of the darkest of the red deer. Very dark brown, nearly black in winter, slightly lighter in summer. Smaller than the Barbary stag and like that animal the antlers lack the bez tine. Found on Sardinia and Corsica.

NORTH AFRICAN RED DEER, BARBARY STAG. Body dark brown, usually with some white spots on the flanks and back. A grayish brown dorsal stripe and a light rump patch. Antlers usually lacking the bez tine. Smaller than the typical red deer. Found in the forested area in northeastern Algeria and western Tunis. A rare animal and the only deer now native to Africa.

NORTH AFRICAN RED DEER *(Cervus elaphus barbarus)*

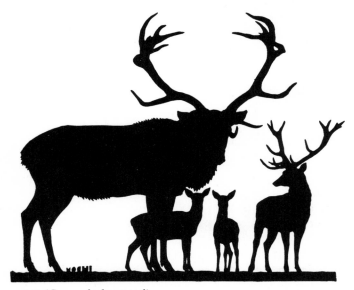

MARAL *(Cervus elaphus maral)*

MARAL, CASPIAN RED DEER. A large heavily built deer, dark slaty-gray with a bright yellow rump patch in winter but general color reddish in summer. Black or dark brown on the shoulders, thighs and underparts. Antlers heavy and massive but less complex than the typical red deer. Found in northern Iran, Crimea and Asia Minor, this deer shows characteristics of both the red deer and the wapiti, and on that account scientists now believe that both these deer should be considered the same species, with the different wapiti simply a subspecies of *Cervus elaphus*. This system of classification is used here except for the New World wapiti. For a number of years the large Asiatic wapiti have been considered a subspecies of the American wapiti and it would rightly follow that these latter wapiti should be included as subspecies of the European red deer. However, this reclassification is too radical a change for a book of this type, and the old name has been retained here.

CAUCASIAN RED DEER *(Cervus elaphus caucasicus)*

CAUCASIAN RED DEER. Closely related to the maral but averaging slightly smaller in size, with lighter, more evenly spreading horns. Confined to the Caucasus.

YARKAND STAG. Light rufous in color with a large light-colored rump patch that includes the tail. Antlers usually five-tined, the terminal fork generally so placed as to point forward. Fifth tine larger than the fourth and inclined inward. Found in eastern Turkestan.

YARKAND STAG *(Cervus elaphus yarkandensis)*

HANGUL *(Cervus elaphus hanglu)*

HANGUL. General color brown with a speckling to the hairs. Light rump patch small, but does not include the tail, in which respect it differs from the Yarkand stag. Inner sides of buttocks grayish white, a line on the inner sides of the thighs and upper side of tail black. Total number of points generally five on each side, the beam being strongly curved inward, the brow and bez tines generally quite close together and at a considerable distance above the burr. Found in the high valleys and mountains of Kashmir.

SHOU. Similar to the hangul but larger, with more massive horns. Beam bending forward at the bez tine so that the upper part of the horns has a forward sweep. General color speckled light brown with a small rump patch that includes the tail. Found in Bhutan, Tibet and west to northern Afghanistan.

BOKHARA DEER, BACTRIAN WAPITI. General color ashy-gray with yellowish sheen, rump patch grayish white, a slightly marked dorsal stripe. Margin of upper lip, lower lip and chin, white. Antlers light in color, normally four-tined, bez tine absent, fourth tine better developed than the third. Found in Russian Turkestan.

22

SHOU *(Cervus elaphus wallichi)*

BOKHARA DEER *(Cervus elaphus bactrianus)*

3
THE WAPITI
AND THOROLD'S DEER

ROCKY MOUNTAIN WAPITI *(Cervus canadensis nelsoni)*

The wapiti, perhaps better known in America as the elk, is, next to our moose, the largest of our deer. In North America the wapiti formerly was found from the mountains of our eastern states west to the Pacific coast. It was exterminated over much of its range and is now restricted to the Rocky Mountain section of the United States and southern Canada. The male, the most stately of our deer, carries a beautiful pair of antlers, which have been recorded up to sixty-six inches along the beam. A typical head has six points on each side but this is often exceeded. By March these antlers are shed and by September they are fully grown again and the bull is in his full glory.

In the Rocky Mountain region the wapiti goes up into the mountains during the spring and summer but the snow forces

it down into the more protected valleys where it often collects in very large herds. Domestic stock and fences have restricted the wapiti's winter range and in many places, as in Jackson Hole, Wyoming, the animals congregate in such numbers that they exhaust the food supply and artificial feeding is carried on by the National Government.

In former years this deer was found in large numbers but unfortunately for the animal its flesh was held in high esteem, its hide also had commercial value and its canine teeth brought fancy prices as watch charms. This heavy demand soon took its toll and the wapiti became extinct over much of its former range. Thanks to government protection the extensive slaughter has ceased and the wapiti numbers have increased. They have been reintroduced in numerous places in the wilder sections in the east and have readily established themselves.

During the mating season, which occurs in the fall, shortly after the antlers are fully developed, the bull wapiti makes his presence known by "bugling." The result of the mating is generally a single spotted calf.

In the humid forest belt of the Pacific slopes in northern California, Oregon, Washington and on Vancouver Island, the largest and darkest form of wapiti occurs, the OLYMPIC or ROOSEVELT'S WAPITI, *C. c. roosevelti.* The antlers of this wapiti are heavy but inclined to be shorter than those of the Rocky Mountain wapiti.

The tule or dwarf wapiti is only slightly smaller than the Rocky Mountain wapiti. Before 1860 these deer were common in the San Joaquin and Sacramento valleys of California, but when these valleys were opened to farming the wapiti were destroyed. Only a few survived and of these the majority were protected on a privately-owned ranch. In 1932 it was estimated that about 170 of these animals existed.

The Asiatic wapiti bear a very close relationship to the American animal and their habits are also similar in that they prefer mountainous districts. They are becoming scarce over most of their range and they, along with other Asiatic deer, are being incessantly hunted by the Chinese for the sake of their antlers which, when in velvet, are ground up for medicine. Each year thousands of these antlers find their way to the Chinese markets.

The rare Thorold's deer inhabits the high plateaus of eastern Tibet and western China. Since the inaccessibility of its habitat makes it very rarely hunted by white men, little is known of its habits. However, it has been so persecuted by the natives for its antlers that it is evidently nearing extinction. This animal appears to prefer the low scrubby underbrush near timber line rather than the forest itself. It goes about in small herds, seldom staying very long in one locality.

TULE WAPITI *(Cervus canadensis nannodes)*

THE ROCKY MOUNTAIN WAPITI. Sides and back brownish gray. A large rump patch straw-colored. Head, neck and legs dark brown, as are the underparts. One of the largest of the deer, with large well-spreading antlers. Found in the Rocky Mountain region from northern New Mexico and Colorado north into Alberta.

TULE OR DWARF WAPITI. The smallest of the New World wapiti. Color much paler than the Rocky Mountain wapiti with more white on the ears. The light rump patch is small and narrow. Restricted to western Kern County, California.

ALTAI WAPITI. Slightly smaller and lighter in color than the Tien-Shan form, and the brow and bez tines rise in close proximity to each other. Found in the Altai Mountains, western Mongolia.

ALTAI WAPITI (*Cervus elaphus asiaticus*)

MANCHURIAN WAPITI (*Cervus elaphus xanthopygus*)

MANCHURIAN WAPITI. General color in summer reddish brown; in winter brownish gray, with dark hairs on the neck and dark underparts. A light-colored rump patch. Antlers shorter and stouter than the Tien-Shan form and the animal smaller. Found in southeastern Siberia, northeastern Mongolia, Manchuria, northern Korea and northeastern China.

TIEN-SHAN WAPITI. The largest of the Asiatic wapiti and quite similar to the American wapiti but the rump patch not so wide and more orange in color. General color brownish gray, the head and neck being darker. Horns long and massive with long tines. Found in the Tien-Shan Mountains of Central Asia.

TIEN-SHAN WAPITI (*Cervus elaphus songaricus*)

THOROLD'S DEER *(Cervus albirostris)*

THOROLD'S OR WHITE-LIPPED DEER. Dull brown above, darker on head, face and neck. A buffy rump patch which includes the tail. Underparts lighter. The coarse hair on the withers grows forward, producing a slight elevation. Inside of ears, chin, upper lip and end of muzzle, white. Antlers somewhat flattened, whitish, the beam tending to turn toward the rear at the third tine, and the bez tine lacking. Found in western China and eastern Tibet.

M'NEILL'S DEER. A large deer of the wapiti type found along the Chinese-Tibetan border. It is of a very pale color finely spotted with gray or brownish black, the winter coat having a brownish wash. This rare deer was first described from a female specimen in 1909, and it was not until twenty-six years later that male specimens with antlers were collected and its true status definitely known.

M'NEILL'S DEER *(Cervus elaphus macneilli)*

FALLOW DEER *(Dama dama)*

MISCELLANEOUS EUROPEAN AND ASIATIC DEER

MESOPOTAMIAN FALLOW DEER *(Dama mesopotamica)*

The history of Père David's Deer is of particular interest since it is unknown in the wild state. In 1865 the French Missionary Père David discovered a herd in a large enclosure in the hunting park of the Chinese Emperor, south of Pekin. A year later specimens were sent to Paris and were described by Milne Edwards. When live animals were shipped later to Europe, the Duke of Bedford fortunately procured some and started a herd in his world-famous deer park at Woburn Abbey. The original herd in China was entirely destroyed in 1900, and the twenty specimens which formed the Woburn herd were, in 1901, believed to be the only ones in existence. Good management had increased this herd to well over 200 by 1935. In 1946, two pairs were brought to America and were deposited in the New York Zoological Park where they have already produced a number of calves. The young, as with most deer, are spotted at birth.

The sika deer is a medium-sized deer which readily adapts itself to domestication. For years it has been kept in the deer parks of Japan and China. It was held sacred by the Japanese; while in China it was bred for its horns which were used for medicine. Eight subspecies are recognized, ranging from southeastern Siberia and Manchuria to southern China. It again occurs one thousand miles south in Vietnam.

BARASINGHA *(Cervus duvauceli)*

FALLOW DEER. Yellowish brown, darker on head and neck, with numerous white spots on the body, but in the winter coat most of these white spots are lost. Underparts white. In semi-domestication the colors vary widely. There is a dark brown variety with little spotting, and white individuals are not rare. Antlers have a well-developed brow tine and long beam, but the antler is palmated at its extremity and from this palm a number of points extend. Originally found in the countries bordering on the Mediterranean from Spain to Iran; introduced into Great Britain and other sections of Europe.

MESOPOTAMIAN FALLOW DEER. Larger than the fallow deer and much brighter in color, with very distinct spotting. Antlers with very small brow tine, broadening in the vicinity of the large trez tine. The extremity of the beam is not palmated but broken up into a number of small tines. Found in the Suristan Mountains in Mesopotamia, Iran and parts of Asia Minor.

BARASINGHA OR SWAMP DEER. General color rufous-brown, darker in the males, sometimes with a speckling of white. Antlers have a long brow tine but no bez or trez, the beam forking about halfway up its length. These forks again fork, giving a cluster of four tines. Found in open forests, plains and swamps in Assam and northern and central India.

THAMIN *(Cervus eldi)*

THAMIN, ELD'S DEER. The mature males are dark brown in color with long coarse hair on the neck. The females are fawn color. The horns have a long, curved brow tine which forms a continuous curve with the beam, making one think of the rocker of a rocking chair. The beam has a number of tines near its end. Typical species from Manipur, but subspecies occur in Laos, Thailand, Hainan, Burma and Tenasserim.

SCHOMBURGK'S DEER. Uniform brown, darkest on nose and upper surface of the tail; cheeks and flanks lighter. Underparts, lower surface of the tail and lower lip are whitish. The beam of the antlers is very short with a long brow tine, which is frequently forked, the beam forking again a short distance above the burr, which fork may again fork. Beyond this first fork the beam may again fork into a number of tines. The whole aspect of the antlers is a broad spread with a flat appearance. Formerly found in a restricted area in Siam. Now believed to be extinct.

SCHOMBURGK'S DEER *(Cervus schomburgki)*

JAPANESE SIKA DEER (*Cervus nippon nippon*)

JAPANESE SIKA DEER. In summer the coat is chestnut-red with numerous white spots, and much browner with no, or only traces of, white spots in winter. Antlers of the sikas are simple, generally with four tines, lacking the bez. Found in Japan and northern China.

FORMOSA SIKA DEER. Slightly larger than the Japanese sika from which it differs in not having the dark winter coat and in retaining its spots the year round. Found on the island of Formosa.

FORMOSA SIKA DEER (*Cervus nippon taiouanus*)

DYBOWSKI'S DEER, PEKIN SIKA. The largest of the sika deer. In color similar to the Japanese sika and may retain some of the spotting in the winter, especially in the females. Found in Manchuria, Korea, southeastern Siberia and northern China.

DYBOWSKI'S DEER (*Cervus nippon hortulorum*)

PÈRE DAVID'S DEER *(Elaphurus davidianus)*

PÈRE DAVID'S DEER, MI-LOU. General color reddish gray, lower part of limbs paler; buttocks, muzzle, area around eye and underparts are whitish. Antlers forking a short distance above burr, the front, more upright prong forking again while the rear prong is long and straight. Ears small, head long, feet large, body long, tail with long blackish brown hairs.

EUROPEAN ROE BUCK. In winter the color is dark brown speckled with buffish yellow, with a large rump patch. In summer the coat is red and the rump patch is wanting. Tail very short. Antlers rise close together and almost vertically from the head with no brow tine but fork about two-thirds from the base, the rear prong again subdividing. The typical roe buck comes from Scandinavia. The European roe buck is found throughout the greater part of Europe.

SIBERIAN ROE BUCK. Larger and paler in color than the European roe buck with the white rump patch larger. Antlers are longer and rougher with numerous snags on the lower section, at times forming antlers of large proportions. Found from the Caucasus, the Altai and the mountains of Turkestan east to eastern Siberia.

OPEAN ROE BUCK *(Capreolus capreolus capreolus)*

37

SIBERIAN ROE BUCK *(Capreolus capreolus pygargus)*

TIEN-SHAN ROE BUCK
(Capreolus capreolus tianschanicus)

MANCHURIAN ROE BUCK
(Capreolus capreolus bedfordi)

TIEN-SHAN ROE BUCK. Similar to the Siberian roe buck but antlers are heavier, spread more widely and are more branched. Found in Tien-Shan Mountains.

MANCHURIAN ROE BUCK. Larger than the European roe buck with the winter coat less brown. Antlers are smaller and more slender. Found in Korea, Manchuria, Mongolia and Shansi Province, China.

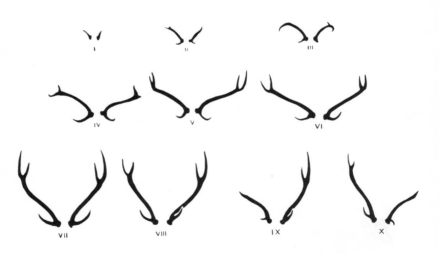

ANTLERS SHED BY AN INDIVIDUAL MALE AXIS DEER
DURING TEN CONSECUTIVE YEARS

AXIS DEER *(Axis axis)*

AXIS DEER, CHITAL. General color light rufous-fawn, much lighter on the head and throat. Black on the muzzle. Profusely spotted with white at all seasons and ages. Horns with brow tine and a terminal fork. Found throughout the open country of India and Ceylon.

HOG DEER. General color dark brown with a yellowish or reddish tinge, the white tips of many of the hairs giving the coat a speckled appearance. In summer the coat often with pale brown or white spots. Underparts paler. The antlers have a brow tine and a simple fork at the extremity of the beam. Found in Assam, Burma, northern India, Siam and Vietnam. Introduced into Ceylon.

PHILIPPINE SPOTTED DEER. General color dark brown, paler on cheeks and occiput. Throat dirty white. Shoulders, back and sides, marked with yellowish white spots, underparts yellowish white. Found on a number of the islands in the Philippines including Cebu, Leyte, Masbate, Negros, Panay and Samar.

HOG DEER *(Axis porcinus)*

PHILIPPINE SPOTTED DEER *(Cervus alfredi)*

INDIAN SAMBAR *(Cervus unicolor unicolor)*

INDIAN SAMBAR. The horns of the sambar and rusa are generally three-tined, consisting of a brow tine and the beam which is forked at its extremity. In the sambars the horns are rough and heavy. Ears large and broad, and the tail bushy. The Indian sambar is the largest of the group, with longer horns and the tines of the terminal fork generally about equal. Color dark umber-brown, with the underparts but slightly paler than the back. Found in India and Ceylon.

MALAY SAMBAR. Slightly smaller than the Indian sambar. Antlers shorter and thicker with the rear or inner tine of the terminal fork shorter than the front tine. Brow tine relatively long. Old males slaty-gray, a light ring around the eye. Limbs lighter than the body in color. Found in

MALAY SAMBAR *(Cervus unicolor equinus)*

Southern China, Hainan, Assam, Burma, Vietnam, Thailand, and south through the Malay Peninsula to Borneo and Sumatra.

FORMOSA SAMBAR. Closely related to the Malay sambar, but the inner sides of the thighs and the whole of the lower legs whitish yellow, and the bushy tail black all around. In winter the general color blackish brown. Antlers smaller than those of the Malay sambar, but of the same general shape. Found on Formosa.

FORMOSA SAMBAR *(Cervus unicolor swinhoei)*

PHILIPPINE SAMBAR *(Cervus unicolor philippina)*

PHILIPPINE SAMBAR. General color rich brown, darkest on back, lightest on neck. Dark brown stripes running from over each eye and meeting to form a band down the center of the face. Found on the island of Luzon. One of the many forms of small sambar deer found in the Philippines.

JAVAN RUSA. The rusas are smaller than the sambars and of a lighter color. The ears are not so large and the horns, though of the same general shape, are lighter in weight and not so rough. The tail not so bushy as the sambars. The Javan rusa has a well-developed mane on the throat and neck. Tail thinner than that of the sambar and ending in a tuft of hairs. General color grizzled grayish brown, browner on the hindquarters and thighs. Underparts dirty yellowish white. Found on Sumatra and Java.

JAVAN RUSA *(Cervus timoriensis tunjuc)*

MOLUCCAN RUSA *(Cervus timoriensis moluccensis)*

MOLUCCAN RUSA. Smaller than the Javan rusa and lacks the distinct mane on the neck, and the tail tuft. General color yellowish brown, the hairs of the back being distinctly marked with alternate dark and light rings, giving a speckled appearance. Found in the Celebes and Moluccas.

JAVAN MUNTJAC. Muntjac are small deer. The antlers are short with unbranched beam and short brow tine, and are supported by long skin-covered pedicles which continue down as convergent ridges on the forehead. The canine teeth are long and sharp, protruding slightly below the lips. The Javan muntjac is deep chestnut-red, paler on the underparts; limbs and face brownish. Found in Java.

INDIAN MUNTJAC. Similar to the Javan muntjac but usually smaller. This race is confined to southern India. The muntjac of northern India is known as *M. m. vaginalis* and is bright chestnut in color. The Burma race, *M. m. grandicornis,* is distinguished by its tawny yellow color and its large horns. Other races are found in southern China.

JAVAN MUNTJAC *(Muntiacus muntjak muntjak)*

INDIAN MUNTJAC *(Muntiacus muntjak aureus)*

TENASSERIM MUNTJAC *(Muntiacus feae)*

HAIRY-FRONTED MUNTJAC *(Muntiacus crinifrons)*

TENASSERIM MUNTJAC. General color dark brown, forehead, region above eyes, ears and pedicle yellow. Black line on the inner side of the pedicle. Horns small. Tall black above and white below. A very rare species confined to Tenasserim, south Burma.

HAIRY-FRONTED MUNTJAC, BLACK MUNTJAC. The largest of the muntjacs and also the rarest. General color dark blackish brown, head lighter brown. Underside of tail white. Well-developed tuft of long light brown hair on forehead and between antlers. Antlers short on a long pedicle. Only four specimens known to exist, all from Chekiang Province, China.

WEST CHINA TUFTED DEER. The largest of the tufted deer and also the darkest. General color chocolate brown, much darker in winter. Underside of tail pure white. Pedicles short and slender; horns very short, curving inward slightly. Tuft of hair on forehead. Upper canines of the males long. Found in western Szechwan Province and eastern Tibet south to border of Burma. Two smaller, grayer subspecies are found in central and southeastern China.

WEST CHINA TUFTED DEER
(Elaphodus cephalophus cephalophus)

44

CHINESE WATER DEER *(Hydropotes inermis)*

CHINESE WATER DEER. Long coarse hairs yellowish brown, slightly brighter on center of back. Underparts white. A small deer with a harsh coat, lacking antlers and having a very short tail. Canines very long in the male. Found in the great marshes of the Yangtze River west as far as Hupeh. The Korean form, *H. i. argyropus,* inhabits that country. The young are spotted and may number as many as five or six in a litter.

MUSK DEER. A small, stockily-built deer with heavy legs, coarse pithy hair, long canine teeth in the male—projecting well below the lips, a very short tail and no antlers. The male has a musk-secreting gland in the skin of the abdomen. General color brownish with reddish-brown and yellow spots, underparts brownish gray. Found from Siberia and Korea south through western China into the Himalayas. A number of subspecies are recognized. Prefers mountain forests and is much persecuted on account of its musk.

MUSK DEER *(Moschus moschiferus moschiferus)*

45

5
THE AMERICAN DEER

NORTHERN WHITE-TAILED DEER *(Odocoileus virginianus borealis)*

The white-tailed deer prefers forest and second growth, and is well able to take care of itself even near towns and cities. It is found from the Atlantic Coast west almost to the Pacific Coast in Oregon, and from southern Canada south into northern South America.

The mule deer is a western deer preferring rough, broken or mountainous country and also inhabits some of the desert regions of the southwest.

The black-tailed deer is a small dark deer which inhabits the coastal forest areas from central California north. In Oregon and California its range extends over onto the eastern slopes of the Cascades. In some of the eastern sections it occupies territory with that of the mule deer, and as these two deer readily interbreed it is now considered a subspecies of that animal. THE SITKA DEER, *O. h. sitkensis,* is a northern race found along the coast and the coastal islands as far north as Juneau, Alaska.

The brockets, small deer preferring forested areas, are found from the Vera Cruz area of Mexico south to Paraguay and Argentina.

Of the remaining South American deer, the guemals prefer mountainous countries, feeding above timberline. The smaller pampas deer prefer the dry open plains, while the large marsh deer inhabit the thick cover of the swamps or the banks of rivers. Most South American deer are becoming scarce.

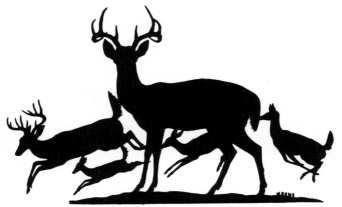

VIRGINIA WHITE-TAILED DEER *(Odocoileus virginianus virginianus)*

NORTHERN WHITE-TAILED DEER. In winter the color is similar to that of the Virginia white-tailed deer, but in summer the pelage is redder. The size averages slightly larger. The antlers usually leave the forehead at a greater angle, giving a slightly flattened spread. Found from southern New England and New York north through southeastern Canada and west to Wisconsin and Minnesota.

VIRGINIA WHITE-TAILED DEER. In winter coat, the body is dark gray mixed with buff, each hair having a black tip. Inner ears, ring about eye, a spot back of nose, upper lip, and chin fringe and under side of tail, white. In summer, the general color of the body is ochraceous-buff. Found from southern New Jersey south to eastern central Florida.

FLORIDA WHITE-TAILED DEER. In color similar to the Virginia white-tailed deer but much smaller in size. Found in southern and western Florida, west along the Gulf Coast to Louisiana.

KEY DEER. This is the rarest and smallest of the eastern deer, the adult male seldom exceeding sixty pounds. It is restricted to the Florida Keys, chiefly Big Pine and neighboring Keys.

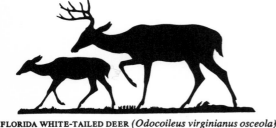

FLORIDA WHITE-TAILED DEER *(Odocoileus virginianus osceola)*

48

KEY DEER
(Odocoileus virginianus calvium)

ARIZONA WHITE-TAILED DEER *(Odocoileus couesi)*

ARIZONA WHITE-TAILED, COUES' DEER. This deer looks like a small light-colored Virginia white-tailed deer with a fine pelage. It is about half the size of the northern white-tailed deer and is found from southwestern New Mexico and southern Arizona south into northern Mexico.

MEXICAN WHITE-TAILED DEER. In Mexico a number of small forms of the white-tailed deer occur. The Mexican white-tailed deer, originally taken from the valley of Mexico, is a typical example.

SAVANNAH DEER. A small white-tailed deer similar in color to the Virginia deer but with small antlers. Found in the savannahs of the Guianas, Venezuela and south to the Amazon River.

XICAN WHITE-TAILED DEER *(Odocoileus mexicanus)*

SAVANNAH DEER *(Odocoileus gymnotis)*

MULE DEER *(Odocoileus hemionus hemionus)*

MULE DEER. Heavier in build than the white-tailed deer, ears much larger and tail rounded and black at the tip. Yellowish brown with large light patch on rump. Dark patch on forehead. Tail white except tip. Dark gray in winter. Antlers double-forked. Found in the Rocky Mountain regions of North America from southern Alaska and Alberta south to Mexico and Lower California.

BLACK-TAILED DEER. In summer, reddish yellow above and brownish gray mottled with black in winter. Tail rather bushy, after the manner of a white-tailed deer but not so long, and upperside black. Underside of tail, chin and upper throat, white. Horns with the upper fork, similar to the mule deer, but occasionally without fork. Found on the Pacific side of the mountains from northern California to British Columbia.

BLACK-TAILED DEER *(Odocoileus hemionus columbianus)*

PAMPAS DEER (*Blastocerus bezoarticus*)

PAMPAS DEER. A small, delicately-built deer of a reddish-brown color with a darker face. Tail blackish brown above, white below. Ring about eye, lips, throat, inner sides of thighs and buttocks, white. Antlers small with a brow tine and a single fork at the end of the main beam. Found on the dry open plains or pampas of Brazil, Paraguay, Uruguay and Argentina.

SOUTH AMERICAN MARSH DEER. Largest of the South American deer. A reddish brown color, brighter in the summer. Legs black below hocks and knees. Underparts white. Horns with a double fork, similar in shape to that of the mule deer. It prefers wooded districts near water. Found from Guiana south through Brazil to northern Argentina.

SOUTH AMERICAN MARSH DEER (*Blastocerus dichotomus*)

PERUVIAN GUEMAL *(Hippocamelus antisensis)*

PERUVIAN GUEMAL, TARUGA. Speckled brown and buffy, coat coarse and brittle. This deer lacks the black face markings of the Chilean guemal. Antlers have but a single fork. Found in the Andes of Ecuador, Peru, Bolivia and northern Chile.

CHILEAN GUEMAL, HUEMUL. Larger than the Peruvian guemal. The coarse hair is a finely-speckled yellowish brown. A broad black band along the nose terminates in a fork between the eyes. Horns generally a simple fork. Tail very short. Found in the mountains of southwestern Argentina and central and southern Chile.

CHILEAN GUEMAL *(Hippocamelus bisulcus)*

BROWN BROCKET (*Mazama simplicicornis*)

BROWN BROCKET. Smaller than the red brocket and of a more slender build. Upper part brown with light brown tips to many of the hairs. Forehead and front of face dark brown. Horns simple spikes. Found south of the Amazon River to northern Argentina. North of the Amazon, a close relative, *M. nemorivaga,* occurs.

RED BROCKET. One of the largest of the brockets. General color bright brownish red; throat, under surface of upper neck and inner sides of thighs, grayish white. Upper side of tail like body; beneath and on tip, white. Antlers simple spikes. Found from Colombia, Venezuela and the Guianas south through Brazil to northern Argentina.

CHILEAN PUDU. Smallest of the deer, of a rich, reddish color. Horns simple and unbranched, generally less than three inches long. Found in the forest regions of south central Chile south along the coast nearly to the Straits of Magellan. Also found on the island of Chiloe. Two local forms occur in northwestern South America, the ECUADORIAN PUDU, *P. mephistophiles mephistophiles* and the COLOMBIAN PUDU, *P. m. wetmorei.*

RED BROCKET (*Mazama americana*)

CHILEAN PUDU (*Pudu pudu*)

6
THE REINDEER AND THE CARIBOU

SCANDINAVIAN REINDEER *(Rangifer tarandus tarandus)*

The Old World reindeer and the New World caribou, which are so closely related that they are often considered subspecies, are well adapted for their Arctic life. Their dense, pithy coat, which is unspotted at all ages, is warm and weatherproof, their hoofs are wide and rounded and the two false hoofs on the rear of each foot are well developed, thus forming a broad surface and enabling the animal to traverse the snow or soft muskeg without difficulty. The antlers are large, situated high on the skull and show considerable palmation. A sharp bend generally occurs in the middle of the beam, behind which there is a back tine. Females generally carry more simple and smaller horns. The ears and tail are short. Their food consists chiefly of lichens, although grass and browse are also eaten. The Barren Ground caribou is noted for its long seasonal migrations.

The reindeer has been domesticated by the Lapps of Lapland for many generations. Life for these people would be most difficult without this animal, for it provides food, clothing and transportation. Even thread is made from the leg sinews. So great was its importance to these people that the Eskimos were in danger of starvation when the wild caribou of northern Alaska became depleted. Domesticated reindeer were imported by the government and have now become well established.

SIBERIAN REINDEER *(Rangifer tarandus sibiricus)*

SCANDINAVIAN REINDEER. General color grayish brown above, neck and underparts whitish. Antlers quite similar to the Barren Ground caribou but averaging less in length and having a distinct back tine. Originally found throughout the alpine region of the Scandinavian Peninsula but now restricted to Norway. The ancestor of the domesticated reindeer.

SIBERIAN REINDEER. Closely related to the Scandinavian reindeer but lighter in color, the color of the body with less contrast to that of the neck and lower parts. Hair longer, especially on the lower neck. Antlers approaching the type of woodland caribou. Found in Siberia and on the tundras of eastern Europe as well as on the Arctic islands to the north.

BARREN GROUND CARIBOU. General color grayish brown with brown on the muzzle and legs. Underparts white or yellowish white. Darker in the summer. The typical Barren Ground caribou is found on the Barren Grounds of North America west of Hudson Bay to the Mackenzie River and northeast to Baffin Land.

BARREN GROUND CARIBOU *(Rangifer arcticus arcticus)*

PEARY'S CARIBOU (*Rangifer arcticus pearyi*)

PEARY'S CARIBOU. The smallest of the caribou, nearly pure white in color, often with a gray area on the back. Their horns tend to have a more upward sweep than those of the typical Barren Ground caribou. Found in Grant Land, Grinnell Land and Ellesmere Land and noted by Peary as far north as latitude 83°.

LABRADOR BARREN GROUND CARIBOU. In color this subspecies is very similar to the Barren Ground caribou. The main difference is in the antlers, those of the present form being characterized by their wide spread and their long sweep, with a backward and then forward curve. The brow and bez tines have great palmation. Found on the Labrador Peninsula north of the forested area.

57

LABRADOR BARREN GROUND CARIBOU (*Rangifer arcticus caboti*)

GREENLAND CARIBOU *(Rangifer arcticus groenlandicus)*

GREENLAND CARIBOU. Darker in color than the typical Barren Ground caribou and slightly larger. Antlers of the males large and spreading, with little palmation in the brow tines. The main beam bends forward at almost right angles. A small rear tine is generally present. Formerly found throughout most of the coastland of Greenland except the northern part. Now disappeared from the east coast but still found on the southwest and west coasts.

STONE'S CARIBOU. A large Barren Ground caribou of dark color with white neck and fringe on the throat, and long, heavy horns. This is the caribou that is found throughout most of Alaska and western Yukon.

STONE'S CARIBOU *(Rangifer arcticus stonei)*

GRANT'S CARIBOU *(Rangifer arcticus granti)*

GRANT'S CARIBOU. Slightly smaller and paler than Stone's caribou; the antlers tend to have a greater spread and more sharply recurving beams, and are lighter in weight. Restricted to the Alaskan Peninsula and Unimak Island.

OSBORN'S CARIBOU. Darker than Stone's caribou; and tending to be slightly larger, with dusky tipping to the white hairs on the throat. When first described this caribou was thought to be of the woodland group on account of the flattened beams of the type. Later it was considered a mountain caribou, but the latest classification identifies it as a Barren Ground caribou. Found in the southern Yukon and northern British Columbia.

OSBORNE'S CARIBOU *(Rangifer arcticus osborni)*

WOODLAND CARIBOU (*Rangifer caribou*)

WOODLAND CARIBOU. General color dark brown, the neck and lips grayish white. The beam of the antlers is shorter and more flattened than the more rounded antlers of the Barren Ground caribou, and the antlers are more compact, extending outward and forward without the long sweep. The brow tine well palmated. Inhabits boreal evergreen forests from Nova Scotia and Quebec west to the Rocky Mountains. Now rare in the east. South of the Saint Lawrence River found only in the Shickshock Mountains of the Gaspé Peninsula. A few may still occur in northern Minnesota, otherwise extinct in the United States. West of the Hudson Bay, the western form *R. c. sylvestris* occurs.

NEWFOUNDLAND CARIBOU. This caribou belongs to the woodland group. It is lighter in color than the mainland form. Antlers much forked, broad, low and widely spreading, the points generally pointing forward and inward with well-palmated brow and bez tines. Found only on the island of Newfoundland.

60

NEWFOUNDLAND CARIBOU (*Rangifer terraenovae*)

MOUNTAIN CARIBOU. A very dark caribou, blackish brown in general color, neck of the males grayish white. Antlers heavy, often extensively palmated. Found in the mountains of British Columbia and Alberta, formerly extending slightly over the United States border into Washington, Idaho and Montana. Now believed to be extinct in the United States.

MOUNTAIN CARIBOU (*Rangifer montanus*)

SCANDINAVIAN ELK (*Alces alces alces*)

7
THE OLD WORLD ELK
AND THE MOOSE

The elk of northern Europe and Asia and the moose of northern North America are very similar in both habit and appearance; in fact, they are so closely connected that some scientists regard them as the same species. However, it is generally believed that enough differences exist to warrant their division into separate species. They are the largest members of the deer family and are distinguished by their long legs, short neck and broad muzzle. The antlers, which are generally broadly palmated, grow from the side on the forehead at right angles to the skull. The antlers of the moose greatly exceed those of the elk in spread. The record, which is that from an Alaskan moose in the American Museum of Natural History, is seventy-seven and five-eighths inches.

It is unfortunate that the name "elk" was given to the wapiti in North America. This animal is more closely related to the European red deer than to the Old World elk. To simplify matters, the American elk should be referred to as the American wapiti.

The moose and the elk both inhabit the northern forests. They are browsers, feeding on the leaves, twigs and bark of trees and bushes. The moose is very fond of certain aquatic plants, and to obtain these it will wade out into the lake, often in water so deep that it is entirely submerged when its head is down gathering the food. When feeding on the leaves and twigs of a small tree, it will often straddle the tree with its forelegs and thus ride it down, so that it may reach the higher leaves.

Through most of the year the moose are solitary, or two or three may be found together. However in winter, if there is a deep snow, a number will "yard up" together, trampling down the snow in a section of good cover. Here they will stay as long as the food supply lasts. In very deep snow a moose is quite helpless and can easily be overtaken by a man on snowshoes. It is often hunted in this way by the forest-inhabiting Indians who are quite dependent on the moose for their winter meat supply.

The moose sheds its antlers in early winter and by spring new ones are beginning to grow. By early fall they have reached their full size and the velvet is shed. The bull helps rid the antlers of the dried velvet by thrashing through the underbrush and attacking small trees and bushes. As soon as the antlers are clean and hard, the bull is in his prime and goes forth looking for rival bulls with which he can test his strength. He is exceedingly noisy and often his thrashing of the bushes and grunts may be heard a long distance. At this time he is also interested in the cows. The hunters in eastern North America take advantage of this and use a birch bark horn to imitate the call of the cow moose and attract the bull. Hunters also attract bulls by striking bushes with a dried shoulder blade of a moose, thus simulating the noise of a rival bull beating the brush with his antlers. In the west, however, the bulls do not seem so responsive to the calls, and they are hunted by tracking or by watching a meadow or other likely feeding place. In the mountains they are stalked because they can be seen from a distance. The moose has acute hearing and a good sense of smell and if much hunted is quite difficult to approach. The Asiatic elk is often hunted with dogs.

Although moose are forest animals, they will occasionally wander. While hunting Rocky Mountain sheep in Alberta, the author was surprised to see a large bull moose come out of a forested draw far below and follow a game trail which passed close to where he was sitting, well above the timber line. The moose continued up the mountain and passed within thirty feet without paying any attention. He was continuously grunting as he walked along the trail. He certainly looked out of his element in the land of the sheep and mountain caribou.

A single calf, very commonly twins, or occasionally triplets, is born during the early summer.

Besides the Alaskan moose and the American moose there is a third subspecies, the SHIRAS' MOOSE, *A. a. shirasi*. This is smaller, slightly lighter in color and found farthest south, in northern Wyoming, Montana and Idaho. It is this moose which is observed by the visitors to the Yellowstone National Park.

SIBERIAN ELK *(Alces alces cameloides)*

AMERICAN MOOSE *(Alces americana americana)*

SCANDINAVIAN ELK. General color blackish brown but sometimes with a yellowish-gray hue; in winter, darker; legs grayish white. The antlers vary greatly in palmation, in some cases being extensively palmated, much after the fashion of the moose, while others lack the palmation and are similar to the antlers of the Siberian elk. Found in the boreal forests of northern Europe and western Siberia.

SIBERIAN ELK. Agreeing in general color and proportions with the Scandinavian elk, but differing from it in that the antlers generally lack palmation. The type specimen has five tines to each antler, three on the rear section and two forward with very little palmation between. The legs are darker in color. This elk is found in the forested country of Siberia south to northern Mongolia and China.

AMERICAN MOOSE. General color dark blackish brown with the face and nose grayer. Legs gray or brownish gray. Antlers well palmated except in the younger males. Found in the forested area from Nova Scotia, Quebec, New Brunswick, northern Maine, New Hampshire, Vermont, west through southern Canada and the northern states to the Rocky Mountains and north to the Yukon.

ALASKAN MOOSE. The largest of the moose, with larger horns. Darker in color than the American moose, with the legs browner. Found in Alaska, intergrading with the American moose in southern Alaska and the Yukon. The best specimens come from the Kenai Peninsula, Alaska.

ALASKAN MOOSE (*Alces americana gigas*)

PRONGHORN ANTELOPE (*Antilocapra americana americana*)

THE PRONGHORN ANTELOPE 8

The pronghorn is placed in a family by itself, mainly on account of its horns. These are true horns which grow over a bony core, the type which is known as hollow horns, and yet it is the only animal with this type of horn which is shed annually and in which the horn bears a prong. The cattle, sheep, goats and true antelopes have this type of horn but bear them throughout life and in case they become broken they are never renewed. The horns are made of material similar to that which forms hoofs and nails. After death this horn sheath is easily removed from the bony core.

The "horns" of deer and their relatives are made up of bone and are known as solid horns or antlers. These are shed and renewed annually. The antlers are shed, generally shortly after the breeding season, breaking off their pedicles or sockets, which are part of the skull. A scar is left and this is quickly covered over with skin. Soon there appear small knobs which are full of blood vessels and velvety in texture. These knobs and the growing antlers are very sensitive. Within, the bony structure is formed and the antler grows rapidly, soon taking shape. It takes a number of months for the antlers of the larger deer to develop but by late summer they have reached their full growth and hardened. The velvet dries and peels off and the solid hard antler remains.

However, the process of renewal of the pronghorn's horns is quite a different procedure. The new horn starts to grow underneath the old horn sheath, and as it develops, it forces the sheath to drop off. A blackish hair-covered skin envelops the horn core but when the horn reaches its full development, this hardens and is lost. The horns of the female are much smaller than those of the male and at times are hardly visible. Male horns have been recorded over twenty inches in length.

In former times pronghorn swarmed the plains in countless thousands, almost equaling the bison in numbers, but like that animal it was ruthlessly killed. It formed one of the main foods for the pioneers going west. The farming of the prairies, fencing the ranges and introducing cattle and sheep so greatly decreased the numbers of the pronghorn that by 1908 it was greatly feared that it might become extinct. Unfortunately it does not take kindly to captivity as does the bison and cannot be penned up in small enclosures. Due to wise restrictions on the killing of these animals and the forming of reserves where the pronghorn had absolute protection, it has made a remarkable comeback, but such protection must continue if the animal is to remain a denizen of our plains.

The pronghorn is considered the swiftest North American mammal and depends on its speed and keen eyesight for protection. It is an animal of the plains or the rolling foothills where it can see long distances and its speed will not be hampered. The pronghorn can jump horizontally for long distances but is not given to leaping objects as is a deer which prefers broken country. It would rather go under or through a fence than over it. It seems to enjoy its powers of speed and will often race along beside a motor car and pass over the road ahead of it. The author has witnessed this same habit in some of the gazelles and other antelopes in Africa. The white hairs of the rump are long and are often suddenly erected, especially if danger is threatening. The flashes so produced may be seen for a remarkably long way and act as a warning signal to other antelopes.

Fawns are born in May or June. Like many of the deer,

but one young is born to a young doe. In the following years she generally has twins. The grayish brown, unspotted coat of the fawn so well matches the vegetation in which it is hidden that the animal is most inconspicuous and it takes a very keen eye to observe it. This is well, for wolves, coyotes and other of the large carnivores, as well as eagles, are constantly in search of just such a morsel. Besides man, the only enemy the adult pronghorn has to contend with is the wolf and the coyote. The antelope has little to fear from a single coyote which it would soon outdistance, but the coyote has learned to run the pronghorn in relays and thus wear it down.

In the southwest and in Mexico the pronghorn is found in the desert regions where water is scarce, and it is believed that it is capable of going without water for long periods. However, it does feed freely on many species of cacti, and as is well known, a number of the cacti store up a water supply for their own use. Undoubtedly, the pronghorn can obtain enough water from these plants to quench its thirst.

Three subspecies of the pronghorn are recognized. The typical race ranges as far south as Mexico where the pale MEXICAN PRONGHORN, *A. a. mexicana,* is found. Farther west on the peninsula of Lower California, a darker race with short rougher horns occurs. This animal has been named *A. a. peninsularis.*

THE PRONGHORN, PRONGHORN ANTELOPE. General color reddish brown, although some of the desert subspecies are a light tan. Two broad bands on the lower neck, the underparts, rump, and sides of head, white or yellowish white. The horns have but a single prong and the ears are long and pointed. Formerly found on the prairies from southern Alberta and Saskatchewan south into northern Mexico and Lower California.

9
THE HARTEBEESTS

BUBAL HARTEBEEST *(Alcelaphus buselaphus buselaphus)*

Hartebeests are one of the most common of the plains animals. They go about in herds, sometimes by themselves but more often mixed with wildebeests, zebras and gazelles. They are exceedingly swift of foot and, as with other plains animals, depend on their speed and watchfulness to protect them from their enemies. One of the common sights of the African plains is a hartebeest atop one of the numerous ant hills gazing out across the landscape.

The hartebeests of the genus *Alcelaphus* are characterized by their elongated heads and by the uniting of the horns at the base on a pedicle which rises from the top of the skull. They stand high at the withers with the back sloping toward the tail. Both sexes have horns.

The bubal hartebeest, the smallest and most northern of this group, is now believed to be extinct. There are occasional reports of this hartebeest having been seen in recent years but none of these reports has been authenticated. The last specimen of which there is record was a female that died in Paris in 1923. In former days this hartebeest ranged across North Africa from Morocco to Egypt. (A report that it at one time occurred in Palestine and Arabia is now discredited.) This is the only instance of a hartebeest found outside of Africa in modern times; although a fossil hartebeest has been found in India. This animal has been rare for many years. One of the first to record it was Marmol who, in 1573, said that in Barbary it occurred in herds of one to two hundred. However, later observers reported herds of only ten or twelve. It was an animal which preferred the rocky country of the plateaus and was not found on the true desert.

The western hartebeest is related to the bubal but is larger and darker in color and has more massive horns. This animal and its close relatives inhabit the plains of Senegal and Guinea east through Nigeria to Tchad and the Central African Republic.

Coke's hartebeest—often referred to as the "kongoni," the Swahili name for the hartebeest—is one of the best known of the hartebeests for it inhabits the plains of southern Kenya and central Tanzania, the country most visited by big game hunters.

Swayne's hartebeest, also known as the Somaliland hartebeest, is closely related to the TORA, *A. b. tora.* This latter animal is slightly larger and lighter in color, the general color being a pale fulvous with dark markings on the chin and tail tip only. There is a light rump patch. The limbs are light in color except the front of the forelegs. The tora is found in western Ethiopia and the neighboring portion of Sennar near the Blue Nile. The darker Swayne's hartebeest lacks the rump patch.

The Cape hartebeest is a strikingly colored animal. At one time it was one of the common animals of Cape Colony, but like all the large game of that district, it was slaughtered in great numbers so that the typical form, found south of the Orange River, is now practically extinct. At the present time there are about one hundred individuals living in semi-domestication in Natal, which are believed to be of the typical form. The northern subspecies, *A. c. selbornei,* found to the north and west, has fared better and is still plentiful in certain sections of Botswana and is receiving protection in the Gemsbok National Park.

The lelwel hartebeest, with its elongated face, high pedicle and upright horns, is an animal with a most grotesque appearance. The typical lelwel from the southern Sudan is characterized by dark markings on the lower part of the limbs. Farther south, in western Kenya, Uganda, northern Ruanda-Urundi and the Bukoba District of northwestern Tanzania, another form is recognized. This is one of the best known of the lelwel type of hartebeests and lacks the dark markings on the legs. It is the JACKSON'S HARTEBEEST, *A. b. jacksoni*. The lelwel from Tchad and the Central African Republic is known as *A. b. modestus*.

The Lichtenstein's hartebeest has the horn-pedicle short and broad, which fact gives the animal a more normal appearance. Personal observation leads this author to believe that this hartebeest is not gregarious as are some of the other species but is generally seen in two's or three's. However, they are said at times to form small herds. They seem to prefer the more open forested land but are also observed on the open plains.

The various hartebeests appear to interbreed readily, especially in the north central part of Africa where their ranges meet, and a number have proved to be hybrids.

WESTERN HARTEBEEST *(Alcelaphus buselaphus major)*

BUBAL HARTEBEEST. Smallest of the hartebeests, standing about forty-four inches at the withers. Uniformly tawny in color with black tail tuft. Formerly found in Morocco, Algeria, Tunisia and east to Egypt.

WESTERN HARTEBEEST. The body color reddish brown, the forelegs with black streaks below the knee. Found in West Africa in Senegal and Guinea.

COKE'S HARTEBEEST. Reddish buff in color. Found on the plains from southern Kenya to central Tanzania.

COKE'S HARTEBEEST *(Alcelaphus buselaphus cokii)*

SWAYNE'S HARTEBEEST (*Alcelaphus buselaphus swaynei*)

SWAYNE'S HARTEBEEST. Reddish brown in color with white tipping to hairs. Forehead, nose, shoulders and upper forelegs black. Found in Somaliland and southern Ethiopia.

CAPE HARTEBEEST. A rich reddish brown with black on face, shoulders and flanks. The lower part of the rump is lighter, in some specimens white or yellowish buff. Formerly ranged through the Transvaal and Orange Free State south to the region of Cape Town. Still found in Botswana and vicinity and common in northern Kalahari Desert.

CAPE HARTEBEEST (*Alcelaphus caama*)

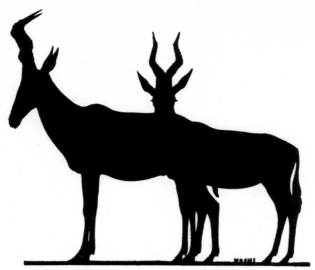

LELWEL HARTEBEEST *(Alcelaphus buselaphus lelwel)*

LELWEL HARTEBEEST. A large dark reddish animal with an exceedingly long face accentuated by the very high pedicle. Found from southern Sudan, southwestern Ethiopia and northeastern Congo south to north-western Uganda and Kenya.

LICHTENSTEIN'S HARTEBEEST. A tawny color with a brighter, more rufous patch on the back. The front of the fore- and hind-limbs black. Found in Mozambique, Malawi, Rhodesia, Zambia to southern Tanzania.

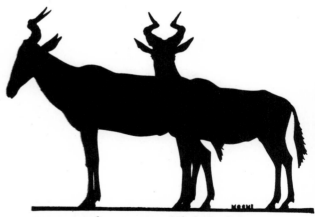

LICHTENSTEIN'S HARTEBEEST *(Alcelaphus lichtensteini)*

80

10
THE DAMALISCUS
HARTEBEESTS

HUNTER'S HARTEBEEST *(Damaliscus hunteri)*

The Hunter's hartebeest, Hunter's antelope or hirola, is so local in distribution that it is not frequently taken by sportsmen unless a special trip is made for that purpose. Like the hartebeest, it prefers the grassy plains and the open bush country, where it associates in small herds of fifteen or twenty animals.

The damaliscus hartebeests are closely related to the true hartebeests but are less specialized. Their heads are less elongated and the horns rise independently from the crown instead of being united at the base.

The Senegal hartebeest, the topi and the tiang are very similar and are now considered subspecies. In habits they resemble the hartebeests, often forming large herds at times intermixed with other game. They are extremely agile and swift of foot and their satin-like coats give them a sleek appearance.

The sassaby, also known as tsessebe, is a more southern representative of the group. It generally is found in small parties of from eight to ten but at certain seasons of the year these groups may unite to form larger herds. In color it resembles its northern relatives but the greater spread of the horns is distinctive.

The bontebok and blesbok are the smallest and most southern members of the genus. They are easily recognized by the white blaze on the face. This blaze is not developed until the animal matures. In the young animals this distinctive mark is black or dark brown.

The bontebok formerly roamed the velds of Cape Colony by the thousands, but it was so persecuted that by 1870 it was at the point of extinction, having been reduced to a single herd. These animals were carefully protected and they have slowly increased. The Bredasdorp Bontebok National Park, in the southernmost tip of Africa, was established for the express purpose of saving this animal from extinction, and in 1938 it was reported that there were sixty-nine bonteboks in the park and a few others on a farm in the vicinity.

The story of the blesbok is very similar to that of the bontebok. Originally living on the plains of the Orange Free State, Botswana and the Transvaal it, too, was greatly reduced in numbers although not so drastically as the bontebok. At the present time it is found only on private farms and ranches and in the Somerville Reserve in the Orange Free State.

SENEGAL HARTEBEEST (*Damaliscus lunatus korrigum*)

TOPI *(Damaliscus lunatus jimela)*

HUNTER'S HARTEBEEST. A uniform tan color with a narrow white band across forehead, the longer hairs of the tail mostly white. It has the long face of the hartebeest but the horns rise from the crown independently as in the Damaliscus hartebeest and are longer and more curved, thus showing characteristics of both groups. Restricted to a small area on the north side of the Tana River in Jubaland, Kenya and southern Somalia.

SENEGAL HARTEBEEST OR KORRIGUM. Bright reddish brown with black markings extending from shoulders to knees and from flanks to hocks. The face is black with black extending outward to form a line below the eye. The coat with a very distinctive sheen. Found on the plains of West Africa from the Lake Chad region to Senegal.

TOPI. Very similar to the Senegal hartebeest but lacks the black stripe below the eye. Found from northern Uganda south through western Kenya and western Tanzania to Lake Tanganyika. The topi which inhabits the coastal region of Kenya and Somalia is known as *D. l. topi.*

TIANG. Similar in markings to the topi but a shade lighter in color and the horns average a greater length. Found in southern Anglo-Egyptian Sudan and as far south as Lake Albert. In Central African Republic and southern Tchad the subspecies *D. l. lyra* occurs.

TIANG *(Damaliscus lunatus tiang)*

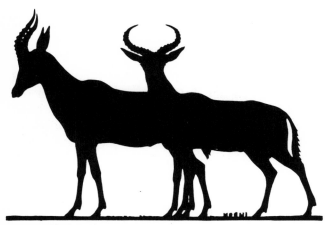

SASSABY *(Damaliscus lunatus lunatus)*

SASSABY. Very similar in coloring and markings to the topi but the horns are more lyre-shaped. Found from Zambia southward to the Orange River.

BONTEBOK. Head, neck, shoulders and flanks a dark reddish brown, the back and upper sides much lighter. Underparts and legs white with narrow brown stripe on forepart of legs. Pure white blaze on face from base of horns to nose, narrowing just above the eyes. The rump and base of tail white, contrasting sharply with the dark flanks. Horns black. Restricted to Bredasdorp National Park.

BONTEBOK *(Damaliscus dorcas dorcas)*

BLESBOK. Similar in color to the bontebok with the reddish brown of the head, neck, shoulders and flanks lighter, causing less contrast with the upper back. The white blaze of the face generally divided by a dark band on lower forehead but occasionally this band is broken as in the bontebok. The color of the back fades on the rump to the base of the tail where there may be a very limited white area. More extensive dark coloring on the lower legs than on the bontebok. The horns of a dark olive green shade.

BLESBOK (*Damaliscus dorcas phillipsi*)

11
THE GNUS

WHITE-TAILED GNU (*Connochoetes gnou*)

The gnus are related to the hartebeests although externally their appearance is very different. The low flowing tail and rounded hips give them a horse-like aspect, but the heavy shoulders and head give them the appearance of diminutive American bison when a herd is seen at a distance.

The popular name for the gnu in Africa is "wildebeest" and this name is becoming more frequently used.

The white-tailed gnu, or black wildebeest, as it is known in South Africa, has much the same history as so many of the larger mammals there. Before the advent of white settlers, large herds of these animals roamed the veld from the Cape as far north as the Transvaal. They were slaughtered in great numbers, and at the present time are found only in a semi-domesticated condition on numerous farms, chiefly in the Orange Free State and southern Transvaal. Their extinction in the wild state is to be regretted, for this gnu received great attention from the early travelers and much has been written about its grace and speed, as well as its sportive capers.

This habit of the black wildebeest is well described by J. C. Millais in his book *A Breath from the Veldt,* written in 1895. He writes:

On meeting with a big troop of Black Wildebeests your approach will have been ignored until within five hundred yards of them—then the animals, if they are lying down, rise and shake themselves and gaze steadily in your direction. When they see that you are still approaching, the whole troop generally commence walking uneasily to and fro, swishing their white tails from side to side with such violence that the whistling caused by their movement can be heard nearly a quarter of a mile away. Then the whole herd, led generally, though not always, by an old cow, prepare to run, affording as a spectacle one of the most curious sights in nature. Not only are the attitudes of the animals themselves queer to the extreme, but in the various formations which a large troop goes through on the veldt are in themselves fascinating to the onlooker. It is a common sight, as the herd stops and faces the hunter, to see a couple of bulls, or even cows, drop on their knees and fight furiously for a minute. Most antelopes, when stopping in the middle of a gallop to survey the intruders, gradually slacken their pace to a walk, and fixing their gaze on him, come slowly to a halt. Not so with the black wildebeest. For when about to halt, the troop may be seen slowly cantering along, till the leader, without stopping a moment to have a look, suddenly turns around, retraces its steps, followed in single file by the whole herd, and not until the very last one has formed up to the new front do they pull up. Black wildebeest never travel very far in their first runs after being alarmed—a herd may be followed and kept at a distance pretty nearly all day—but then they are such excellent judges of distance that you can seldom get nearer than five hundred yards. When trotting fast, considering the size of the animal the stride is immense, and the pace equal to that of most antelopes when galloping.

The brindled gnu or blue wildebeest was more fortunate than the white-tailed gnu and is still common over much of its former range. It prefers the more open plains where it forms large herds, at times associating with hartebeests, zebras and other game. The white-bearded gnu or wildebeest is a more northern form with similar habits.

An experience which stands out most vividly occurred the first morning the author stood on the rim of the crater of Ngorongora, in northern Tanzania. Almost a mile below, the crater floor of this extinct volcano was a vast level plain, some ten miles across, containing a shallow lake and a few scattered low forests. Spread over this plain as far as the eye could see, were thousands of head of game. With glasses we were able to distinguish many of the animals. There were a large number of zebras, some gazelles, about thirty elands, four rhinoceroses; but the vast majority of the animals were white-bearded wildebeests. In places they were so closely herded that the plain was black with them. After breakfast we descended to the crater floor and enjoyed a most interesting day. As this is a reserve, the animals were fairly tame and we had no trouble in securing some interesting pictures.

On the plains of East Africa, zebra, hartebeests and wildebeests form the main food for the lion. In central Tanzania, where wildebeests are common, they comprised a large number of the kills we discovered.

BRINDLED GNU *(Connochoetes taurinus taurinus)*

WHITE-TAILED GNU. Dark brown in color with a white tail which is long and flowing. A tuft of black hairs rises just above the nose and is directed upward and forward. The ends of the hairs of the mane and beard are also black. Now found only in a semi-domesticated state on farms in the southern Transvaal and the Orange Free State.

BRINDLED GNU. Blue-roan in color with dark brown and black vertical lines on the sides of the neck and body. The face, mane, beard and the long hairs of the tail are black. From the Orange River it ranges north into Botswana, northern and eastern Transvaal, Angola, Zambia, Rhodesia, Mozambique and Malawi.

WHITE-BEARDED GNU. Similar in general coloration to the brindled gnu but averaging lighter with the long hairs on the throat, white or yellowish white. Found on the plains of East Africa from central Tanzania to central Kenya.

WHITE-BEARDED GNU *(Connochoetes taurinus albojabatus)*

92

GRAY DUIKER *(Sylvicapra grimmia grimmia)*

12
THE SMALL
AFRICAN ANTELOPES

ATHI GRAY DUIKER *(Sylvicapra grimmia hindei)*

CROWNED DUIKER
(*Sylvicapra grimmia coronata*)

BAY DUIKER (*Cephalophus dorsalis dorsalis*)

There are over eighty forms of duikers now recognized.

GRAY OR BUSH DUIKER. General color grayish brown, more or less grizzled. Dark brown or black patch on upper nose sometimes extending up to forehead, with a line of the same color down the front of the forelegs to hoofs. Chin, inner side of forelegs, thighs and under side of tail, white. Found in open country from the Cape to Ethiopia and west to Senegal.

ATHI GRAY DUIKER. Similar to the gray duiker but of brighter color with less black grizzling. Southern Kenya south through Tanzania.

CROWNED DUIKER. A relative of the gray or bush duikers but of a yellow color finely grizzled with black. A tuft of brown hair extending above forehead. Confined to West Africa.

BAY DUIKER. Bright chestnut-rufous with a dark brown or black dorsal line running from the nose to the tip of the tail. Legs brown. Found from Sierra Leone to Ghana south to Angola.

CHESTNUT DUIKER. Very similar in color to the bay duiker but slightly larger. Color deep chestnut except for broad black dorsal line. Lower part of legs brown. Conspicuous white spots over eyes and on each side of nose. Cameroun to northeastern Congo.

WEYNS'S DUIKER. Rich yellowish red, much brighter on the rump. Bright chestnut on forehead and crest. Browner on neck and legs. No dorsal band. Sharply defined area on nape where the hair is directed forward. Found in the forested regions of northeastern Congo to southwestern Uganda.

CHESTNUT DUIKER (*Cephalophus dorsalis castaneus*)

WEYNS'S DUIKER (*Cephalophus natalensis weynsi*)

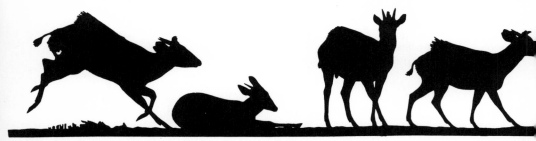

YELLOW-BACKED DUIKER *(Cephalophus silvicultor silvicultor)*

YELLOW-BACKED DUIKER. One of the largest of the duikers. General color blackish brown, much lighter on nose, lower face and throat. A more or less intense yellow dorsal stripe from middle of back broadening toward a grayish rump. Liberia to Angola east to Zambia.

ABBOTT'S DUIKER. Color uniform chestnut-brown becoming darker toward the rump. Crown with long tuft of blackish hairs. Found only in the forests of Kilimanjaro and the Usambara Mountains, Tanzania.

GABOON DUIKER. Buffy brown in color, being much brighter on rump and flanks and darkening at shoulders. A broad blackish brown dorsal stripe from nape to rump extending in a narrow line along upper side of tail. Tail ending in a rounded tuft of black and white hairs. Found from Gabon to the eastern Congo.

ABBOTT'S DUIKER *(Cephalophus spadix)*

GABOON DUIKER *(Cephalophus leucogaster)*

ZEBRA ANTELOPE *(Cephalophus zebra)*

ZEBRA ANTELOPE. Color rich orange brown with a series of black stripes varying to an inch wide, extending over the back. Found in the interior of Liberia and Sierra Leone, West Africa.

MAXWELL'S BLUE DUIKER
(Cephalophus maxwelli)

EQUATORIAL BLUE DUIKER
(Cephalophus monticola aequatorialis)

MAXWELL'S BLUE DUIKER. Color slaty-brown, slightly paler below. Legs similar in color to body. West coast of Africa from Senegal to Liberia.

EQUATORIAL BLUE DUIKER. Slaty-brown becoming dark brown on the rump. Underparts and legs slightly lighter than the back. Found in the northeastern Congo, Uganda and northwestern Tanzania.

SIMPSON'S DUIKER. A small blue duiker with a brownish cast throughout. Russet on the limbs and flanks. Central Congo.

UASIN GISHU ORIBI. Uniform tawny yellow in color. Tufts of hair at the knees and a bare spot below the ears as in the reedbuck. Rudimentary tail. Female slightly larger than male and lacks horns. Oribis are found on the plains from the Cape to Ethiopia and west to Senegal. This subspecies occurs from southeast of Lake Victoria north to central Uganda.

SIMPSON'S DUIKER
(Cephalophus monticola simpsoni)

UASIN GISHU ORIBI *(Ourebia ourebi cottoni)*

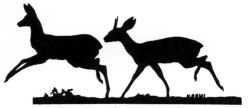

MASAILAND STEINBOK *(Raphicerus campestris neumanni)*

GRYSBOK *(Raphicerus melanotis)*

MASAILAND STEINBOK. Bright sorrel-red color. Tail very short. Ears long. False hoofs lacking. Horns short, rising vertically from the head. Tanzania north to northern Kenya.

GRYSBOK. The hairs of the body long and red, interspersed with pure white hairs giving a grizzled appearance. False hoofs present. South Africa from the Cape north to the Transvaal and Zambia.

MARSABIT KLIPSPRINGER. Hair coarse and pithy. Bright buff-yellow with a speckled appearance. Hoofs rounded, the animal standing on the tips. Female generally hornless. Found throughout most of Africa where there is a sufficient outcropping of rocks. This particular form is found from northern Kenya north to Lake Rudolph and west to Mt. Elgon.

MASAILAND KLIPSPRINGER. Differs only slightly in color from the other klipspringers but the female carries horns. Found only in rocky regions from central Tanzania north to Mt. Kenya.

98

MARSABIT KLIPSPRINGER
(Oreotragus oreotragus aureus)

MASAILAND KLIPSPRINGER
(Oreotragus oreotragus schillingsi)

ROYAL PYGMY ANTELOPE
(*Neotragus pygmaeus*)

SUNI (*Nesotragus moschatus moschatus*)

ROYAL PYGMY ANTELOPE. General color bright rufous-fawn, brighter toward the rump. Chin and underparts white. The smallest of the antelope. Height about ten inches. Found from Sierra Leone south to Ghana. A close relative, the DWARF ANTELOPE, *Neotragus batesi* occurs in Cameroun and eastern Congo.

SUNI, ZANZIBAR ANTELOPE. General color brown with white underparts. Height about thirteen inches. The typical form originally came from two small islands, Bawe and Grave, off the coast of Zanzibar, and also from the adjoining mainland. Subspecies also occur in the Mt. Kilimanjaro region, the highlands in Kenya from Nairobi north to Mt. Kenya and along the east coast from Tanzania north to the Tana River.

PHILLIP'S DIK-DIK (*Madoqua phillipsi phillipsi*)

BEIRA (*Dorcatragus megalotis*)

BEIRA. General color purplish gray grizzled with white. Area around eyes white. Ears large, dwarfing the horns. Restricted to Somalia and locally in eastern Ethiopia.

PHILLIP'S DIK-DIK. Back grizzled with a rufous wash more strongly on the dorsal area. Neck and flanks gray. Lower sides rufous. Nose only slightly lengthened. Northern Somalia and southeastern Ethiopia.

LONG-SNOUTED DIK-DIK. Color more subdued than Phillip's Dik-dik, slightly larger, and proboscis much elongated. Somalia. Ethiopia to Tanzania. KIRK'S DIK-DIK (*Madoqua kirki*), with a proboscis smaller and narrower, has a discontinuous distribution between eastern and southwestern Africa.

LONG-SNOUTED DIK-DIK
(*Madoqua guentheri guentheri*)

COMMON WATERBUCK *(Kobus ellipsiprymnus ellipsiprymnus)*

13
THE WATERBUCKS
AND THE REEDBUCKS

The waterbucks and the reedbucks comprise the subfamily Reduncinae in which the horns are ridged and generally curved although they vary greatly in shape and size. The female does not carry horns.

The waterbucks are characterized by their long coarse hair. Both of the waterbucks agree in habits. The defassa or sing-sing waterbuck is slightly larger than the common waterbuck, and although it averages a more reddish tone, some of the forms such as CRAWSHAY'S WATERBUCK, *K. d. crawshayi* of Zambia, are of a slaty color.

Contrary to the implication of its name, the waterbuck is not so partial to water as the lechwe. Although at times inhabiting swamps and marshes, it may often be found on the dry plains a number of miles from water.

The elongated hoofs of the lechwes distinguish a water-loving antelope. This animal goes about in large herds and prefers the swamps and grass flats along the river. The short nose, long lyre-shaped horns, coarse hair and long tail are characteristic.

The black lechwe has been greatly reduced in number and is now very local. It was the author's good fortune to visit the home of the last herds on the eastern shore of Lake Bang-weulu. We estimated that we saw about six hundred head in four herds.

The Nile lechwe, sometimes known as Mrs. Gray's water-buck, has much the same habits as its southern relatives. This striking animal is so highly prized as a trophy that a law was passed permitting a hunter to take but two animals in a lifetime.

The kob differs from the lechwe in having a longer nose, shorter horns, and a smoother coat. The tail is shorter in the kob, not reaching to the hocks as it does in the lechwe. The kob prefers the higher, dry, open flats in which to graze, and is either found singly or associated in small herds.

The reedbuck reminds one very much of the North American white-tailed deer in size and color as well as in its habit of erecting its bushy tail, thus exposing the pure white lower surface, when running away. It differs from most antelopes in having a bare patch below the ear. It prefers a region close to the rivers or swamps where there are reed beds in which to hide, and where it may graze on the grass in the nearby open country, generally singly or in a party of three or four. When startled, the male often gives forth a shrill whistle. The common reedbuck, known as the rooi rhebok in South Africa, is replaced in the north by the smaller bohor with shorter and more hooked horns. Numerous subspecies of these two animals have been described. As the name implies, the mountain reedbuck is not so partial to water but prefers stony mountainous districts. It also associates in larger herds than the common reedbuck. CHANLER'S REEDBUCK, *R. f. chanleri* is the subspecies found in Kenya.

The vaal rhebok differs chiefly from the true reedbucks in the form of its horns and in lacking the bare patch below the ears. This animal prefers the rocky, open country of the mountains. It is generally found in small herds of from six to ten individuals.

DEFASSA WATERBUCK (*Kobus defassa defassa*)

COMMON WATERBUCK. With less of the reddish tinge of the defassa waterbuck and with an elliptical white stripe extending along the sides of the rump, completely surrounding the tail. Found from the Limpopo River in the Transvaal north through eastern Africa and Kenya.

DEFASSA WATERBUCK. A dark reddish brown in color. Lower legs darker brown. White rump patch. Found from the Zambesi River to Lake Nyasa north into Angola and through the central and western parts of Africa to Gambia and Senegal, the Sudan and western Ethiopia.

COMMON WATERBUCK FEMALES AND CALVES

RED LECHWE (*Kobus leche leche*)

RED LECHWE. Yellowish red in color, the forelegs and often the chest black or dark brown. Found from northern Botswana north to Zambia.

BLACK LECHWE. Adult males are varying shades of blackish brown. Females and young males reddish brown. Found only on the marshes along the eastern and southern shores on Lake Bangweulu, in western Zambia.

NILE LECHWE, MRS. GRAY'S WATERBUCK. Adult male blackish brown with white patch on the shoulders, white around the eyes and white on the head. Female and young male chestnut. Found only on the marshes in the Bahr-el-Ghazel district of the southern Sudan.

BLACK LECHWE (*Kobus leche smithemani*)

NILE LECHWE (*Kobus megaceros*)

WHITE-EARED KOB (*Kobus kob leucotis*)

WHITE-EARED KOB. Old male generally blackish brown in color. Young male and female similar in color to the other kobs except for the white ears. Found in the Sudan in the drainage of the White Nile, the Sabot and the Bahr-el-Ghazel.

BUFFON'S KOB. Reddish yellow in color with black on the front of the forelegs. A white line over the eye. Found from Central African Republic and Tchad west to Senegal and Gambia.

BUFFON'S KOB (*Kobus kob kob*)

UGANDA KOB (*Kobus kob thomasi*)

UGANDA KOB. Slightly larger than the Buffon's kob and of similar color but the white near the eye forming a complete ring around it. Found in the country south of Lake Victoria north to the northern border of Uganda.

PUKU. A small kob of similar color to Buffon's kob but lacking the black stripe on the forelegs. The hair on the back has a tendency to curl. Horns heavier. Found from the Chobi and Zambesi rivers north through Zambia and Malawi.

COMMON REEDBUCK. Largest of the reedbuck. Color, grayish fawn grizzled with brown. South Africa, north to Angola, Zambia and Malawi.

PUKU (*Kobus vardoni vardoni*)

COMMON REEDBUCK (*Redunca arundinum arundiu*

BOHOR REEDBUCK *(Redunca redunca redunca)*

BOHOR REEDBUCK. Color bright fawn. Tips of the horns curving sharply forward. Found from Tanganyika north into Ethiopia and the Sudan, and west to Senegal, Gambia and Ghana.

MOUNTAIN REEDBUCK. Smallest of the reedbuck, of a grayish fawn color with small horns. Eastern Africa from Cape Colony to Ethiopia.

VAAL RHEBOK. Pale gray with somewhat woolly hair. Ears long and narrow. Horns straight and upright. South Africa south of the Zambesi River.

MOUNTAIN REEDBUCK
(Redunca fulvorufula fulvorufula)

VAAL RHEBOK *(Pelea capreolus)*

14
THE GAZELLES
AND THEIR ALLIES

DIBATAG (*Ammodorcas clarkei*)

GERENUK *(Litocranius walleri)*

About sixty forms of true gazelles are recognized.

DIBATAG OR CLARK'S GAZELLE. Deep cinnamon in color with the face markings of a gazelle. Exceptionally long neck, tail and legs. The short, hooked horns are similar in shape to a reedbuck's and at one time it was thought to be related. Female without horns. Confined to Somalia and eastern Ethiopia.

GERENUK OR WALLER'S GAZELLE. Exceedingly long neck and legs with small head. Cinnamon-red in color, darker along the upper back. Horns lyrate and hooked forward at tip and found only on the male. From Somalia and Ethiopia south to Tanzania.

IMPALLA. Cinnamon-rufous in color, lighter on sides. Underparts white. Black stripe on rump. False hoofs absent. Two oval black patches of long stiff hairs on hind legs. Female without horns. Found from South Africa north of the Orange River northwest to Angola and in East Africa to southern Uganda.

109

IMPALLA *(Aepyceros melampus melampus)*

SPRINGBUCK (*Antidorcas marsupialis marsupialis*)

ADDRA GAZELLE (*Gazella dama ruficollis*)

SPRINGBUCK, OR SPRINGBOK. General color cinnamon-buff with a dorsal line of long white hair extending from center of back to white of rump. Dark brown stripe on sides, separating the color of the upper back from the pure white underparts. Both sexes carry horns. Formerly thousands ranged the plains of South Africa. Now greatly reduced. South Africa as far north as Angola in the west and northern Transvaal in the east.

ADDRA GAZELLE. The largest of the true gazelles, with long neck and legs. Body white with a more or less rufous wash varying with the different forms. Head and neck rufous. Deserts of North Africa from Senegal to the Sudan.

DORCAS GAZELLE. Pale fawn with a dark lateral stripe, which varies in intensity with the different subspecies. Distinct face markings. This little gazelle has a very extensive range, from Nigeria and the Sudan to Morocco, Algeria, Ethiopia and Egypt. A close relative is found in Palestine and Syria.

MUSCAT GAZELLE. Similar to the Arabian gazelle but smaller with the tips of the horns turning in. Eastern Arabia.

DORCAS GAZELLE (*Gazella dorcas dorcas*)

MUSCAT GAZELLE (*Gazella muscatensis*)

SPEKE'S GAZELLE *(Gazella spekei)*

LODER'S GAZELLE *(Gazella leptoceros loderi)*

SPEKE'S GAZELLE. Hair longer than that of many gazelles. General color pale brownish fawn, face stripe slightly darker. A black nose patch in front of which there is a fleshy protuberance. Flank band brownish black. Horns with a backward curve, curving in at the tips. Somalia.

LODER'S GAZELLE. Very pale in color with the gazelle head markings indistinctly defined. Deserts of Algeria and Tunis to Nubia and Sennar.

SOEMMERRING'S GAZELLE. A large gazelle of a light cinnamon color with the white of the rump patch extending well forward. The face markings are black and the horns hook inward at the tips. Found in eastern Ethiopia and Somalia.

RED-FRONTED GAZELLE. Very similar to the Thomson's gazelle but differs in having a broad buffy flank band below which is a narrow black one. From Senegal and northern Nigeria east to the Sudan and northern Uganda.

SOEMMERRING'S GAZELLE *(Gazella soemmerringii)*

RED-FRONTED GAZELLE *(Gazella rufifrons)*

111

GRANT'S GAZELLE (*Gazella grant*

THOMSON'S GAZELLE (*Gazella thomsonii*)

THOMSON'S GAZELLE. This small gazelle of the East African plains, locally known as the "Tommy," is of a light rufous color with a black band along the lower side. Underparts white. From central Tanzania north through the higher plains of Kenya.

ARABIAN GAZELLE. Grayish fawn in color with a darker flank band. Horns with a slight backward curve, the tips inclined to point forward. Western and southern Arabia.

EDMI OR ATLAS GAZELLE. General color dull fawn with an indistinct lateral band and well-developed face markings. A blackish spot on muzzle and a rough coat. Horns upright and nearly parallel with a slight backward curve and a forward curve at the tips. Mountains of Morocco, Algeria and Tunisia.

INDIAN GAZELLE OR CHINKARA. Light chestnut. The tips of the horns do not turn in. The white of the underparts does not extend to the root of the tail. The horns of the female are small and sometimes wanting. Plains of India.

ARABIAN GAZELLE (*Gazella arabica*)

EDMI GAZELLE (*Gazella cuvieri*)

INDIAN GAZELLE (*Gazella benner*

ROBERT'S GAZELLE *(Gazella granti robertsi)*

GRANT'S GAZELLE. ROBERT'S GAZELLE. A large gazelle with fawn-colored back, white underparts and white rump patch. The horns are the longest of any of the gazelles and vary in spread with the different subspecies. Robert's gazelle has the greatest spread. From southern Ethiopia south through East Africa to central Tanzania. Robert's gazelle ranges from central Tanzania north in the vicinity of Victoria Nyanza to southern Kenya.

GOITERED GAZELLE. General color pale sandy with indistinct side band. In winter the color is paler and the coat much thicker. Female either lacking or with poorly developed horns. Derives its name from the swelling of the larynx in the male during the breeding season. From northwest Iran and Asia Minor east to the southern Gobi Desert.

ZEREN OR MONGOLIAN GAZELLE. Larger than the Tibetan gazelle. In summer, bright ochraceous-buff. In winter, hair longer and more pinkish buff. Lacks the gazelle head markings. The tail is short but longer than in the Tibetan gazelle. Horns at first nearly parallel then forming an outward curve inclining inward at the tips. Female without horns. Found on the grassy plains of northern Mongolia from the eastern border of that country west to the Altai Mountains and south into Inner Mongolia.

GOITERED GAZELLE *(Gazella subgutturosa)*

ZEREN *(Procapra gutturosa)*

113

GOA *(Procapra picticaudata picticaudata)*

GOA OR TIBETAN GAZELLE. Characterized by the very short tail, tufts of hair on knees, and the lack of horns in the female. The horns of the male are heavily ridged, rise vertically and then curve backward, the tips bending upward and slightly inward. Coat long in winter and pale fawn with large white rump patch. Tibetan plateau, east to western border of China.

PRZEWALSKI'S GAZELLE. Similar to the goa but slightly larger with the tips of the horns turning inward to a greater extent. In summer, dark fawn; in winter, hair longer and more grizzled. Southern Mongolia westward to northern central Kansu.

PRZEWALSKI'S GAZELLE *(Procapra picticaudata przewals*

15
THE SABLES, THE ROANS, AND THE ORYXES

SABLE ANTELOPE (*Hippotragus niger*)

The sable antelope is a superb animal with upstanding carriage and long sickle-shaped horns, and its habit of running with its neck arched—practiced by few antelopes—gives it every appearance of a thoroughbred. Sable prefer open forested country where they go about in small herds, ranging up to about thirty animals. These herds are made up of cows and young bulls with but one dark mature bull in the company. The other bulls lead a solitary life, or two or three may run together. On occasion an old cow will take on the dark color of a bull.

The giant or Angolan sable is similar in color and habits to the sable but the long horns are characteristic. For example, the record length of horn for the giant sable is sixty-four inches against fifty-two and one-half inches for the common sable. This animal is considered such a prize by sportsmen that it is becoming rare and is now afforded protection. It was not until 1916 that it was described.

The roan is one of the largest antelopes, being second only to the eland in size. The sickle-shaped horns are but little longer than the head. The hair is rather coarse and has none of the sleek appearance of a sable. It is an animal of the plains and thornbush country and is found at a considerable elevation on some of the mountains. Like the sable it can become a dangerous animal when wounded.

The bluebok or blaauwbok of South Africa was about forty-five inches at the shoulder and suggested a small roan antelope in appearance. Living in a very limited territory southeast of Cape Town, it did not long survive the arrival of white settlers. As far as is known, the last specimen was killed in 1799 and but five mounted specimens are in existence today, all of them in European museums.

The gemsbok prefers desert country and is especially common on the Kalahari Desert where it goes about in considerable herds, although at times a single animal or a group of three or four occurs. Since rain does not fall for months at a time in many places in its home, this animal is forced to secure what water it needs from the succulent food it eats. Thus it remains sleek and fat. The horns of the female, although slimmer, are often longer than those of the male.

The beisa oryx is the oryx most often taken by sportsmen as it is common in some sections of Kenya. There have been a number of subspecies described, one of which is the FRINGED-EARED BEISA, *O. g. callotis* of Tanzania. This animal differs in having long hair on its ear tips.

The Arabian oryx or Beatrix antelope, is the only member of the oryx group found outside of Africa. Found in the great sandy wastes, it goes without drinking water for most of the year and lives among the most arid desert conditions.

The white oryx and the addax are true desert animals living in small herds, days from water. Like many of the desert animals their hoofs are very broad to help them traverse the deep sands. Their range and habits are very similar.

GIANT SABLE ANTELOPE *(Hippotragus niger variani)*

SABLE ANTELOPE. Old bull very dark brownish black with contrasting white underparts. Female and young bull reddish brown. Occurs from Transvaal north to Kenya where it is found only along the southeastern coast near Mombasa.

GIANT SABLE ANTELOPE. Slightly larger than the common sable, with much longer horns. Found locally in the interior of Angola.

ROAN ANTELOPE. Color red-roan. Ears long and narrow, in some of the subspecies having a chestnut tassel at their extremity. Found from the Sudan, Ethiopia and the Lake Chad region south to the Orange River in South Africa.

ROAN ANTELOPE *(Hippotragus equinus)*

BLUEBOK *(Hippotragus leucophoeus)*

BLUEBOK. Similar in appearance to a small roan, of a bluish gray color, lacking head markings. Formerly existed in the southernmost part of South Africa, southeast of Cape Town.

GEMSBOK. A grayish fawn color. Black or dark brown head markings. Black on lower flanks, a line along the lower part of side and on forelegs to knee. A tuft of black hair on throat, especially noticeable in males. Found in the desert regions of southwest Africa and as far north as southern Angola.

BEISA ORYX. Similar to the gemsbok but smaller and lacks the dark patch on flanks, and the throat tuft. Found from Nubia in the Sudan south through the coastal plains of East Africa to central Tanzania.

GEMSBOK *(Oryx gazella gazella)*

120

BEISA ORYX *(Oryx gazella beisa)*

ARABIAN ORYX (*Oryx leucoryx*)

ARABIAN ORYX. The smallest of the oryx. White with dark chocolate legs. Black blaze on the face, and the black cheek markings are typical oryx. Found only in two sections of Arabia, seven hundred miles apart, in northern Nafud and the Rub al Khali in the southeast.

WHITE ORYX. White, washed more or less with chestnut, becoming darker and more constant on the neck and shoulders. The typical oryx marking on the head, of this chestnut color. Horns forming a long sweeping curve. Found in the desert regions of North Africa from Nigeria to the Sudan.

ADDAX. Brownish gray in color. A patch of slightly longer black hairs on the forehead. A white patch below the eye extends in a narrow line across the front of the face. Found in the desert regions of North Africa and has much the same range as the white oryx.

WHITE ORYX (*Oryx dammah*)

ADDAX (*Addax nasomaculatus*)

THE BLACKBUCK, THE CHOUSINGHA
THE SAIGA, THE CHIR

BLACKBUCK (*Antilope cervicapra*)

The blackbuck is the most common of the game animals of India. It has a very extensive range, being found in the plains region from Cape Comosin, the southernmost tip of India, north to the foothills of the Himalayas, east to southern Assam and west to the Punjab. The long, twisted horns may reach a length over thirty inches. The blackbuck usually go about in small herds but sometimes herds of several hundred may be found. It is an animal of the plains or open forest and is a

grazer. Like most of the plains-dwelling antelopes, black-bucks depend upon their eyes to detect danger as their senses of smell and hearing are not so acute as those of the forest antelopes. The blackbuck is very swift of foot and its stamina is remarkable, for it can continue a fast pace for a number of miles. The cheeta is sometimes used in blackbuck hunting and for a short distance the cheeta, which is considered the fastest animal afoot, can outrun the buck, but if the cheeta does not overtake the blackbuck in the first two hundred yards or so, the buck will easily pull away from its pursuer.

Blackbuck breed throughout the year and single fawns or twins may be born. The mother hides them in the grass, but in a very short time they are able to join and keep up with the herd.

The four-horned antelope is found in open forests where the light permits good grass to grow. It is never found far from water, and unlike the blackbuck, does not make the open plains its home. It generally goes about singly or in pairs, or three or four may be found together, but it is never found in herds. In habitat, habits and appearance, it is similar to the African duikers and at one time was thought to be related. The rear or main horns are recorded up to four and three-fourths inches and the forehorns about half that length. The female is hornless.

The saiga is unique in having a large swollen nose with the nostrils directed downward, very much after the manner of the long-snouted dik-dik. Another peculiarity of this animal is its lyrate, amber-colored horns, the cause of the downfall of the saiga, for the Chinese considered them good medicine and paid high prices for them. Thousands were killed so that the number and range of the saiga were drastically reduced. Only a small remnant of the former thousands survives. It is

124

now afforded some protection, but numbers are still illegally killed. The saiga was formerly found in Poland and other parts of eastern Europe. Saiga generally go about in small bands, but during certain seasons of the year these bands join forces and form extensive herds. On account of the bleak country in which they are found, their winter coat is thick and soft. There is some seasonal migration. The female, which is without horns, is said to have two young as a rule but sometimes a single one.

The chiru also has a nose that is swollen but this swelling is at the sides and not over the top as in the case of the saiga. The black horns are rather straight, curving outward for about two-thirds their length and then inward and have a recorded length of twenty-seven and three-fourths inches.

The home of the chiru is the bleak highlands of the Tibetan Plateau. Brigadier General Alexander Kimloch, who in 1861 and 1862 hunted this animal in the Chang-Chenms valley, writes the following in his *Large Game Shooting in Tibet, the Himalayas, Northern and Central India*:

So far as we know, Tibetan Antelope are never found near the habitations of man, but frequent the plains and elevated valleys far above the limits of cultivation, where few human beings, save occasional wandering shepherds, ever disturb them.

Although living in such remote and sequestered regions the Tibetan antelope is wary in its habits. In the mornings and evenings it frequents the grassy margins of glacial streams which frequently flow between steep banks—these ravines have for the most part been cut through gently sloping valleys and on ascending their steep sides slightly undulating plains will be found to stretch away, until they merge in the easy slopes of the rounded hills which bound the valleys.

To these plains the antelope betake themselves during the day and there they excavate hollows deep enough to conceal their bodies, from which, themselves unperceived, they can detect any threatening danger at a great distance.

The chiru generally goes about in small herds, although occasionally large numbers congregate. Swift of foot, this antelope holds its horns erect while running.

CHOUSINGHA *(Tetracerus quadricornis)*

BLACKBUCK OR INDIAN ANTELOPE. Old bucks are dark brown or black above, lighter on the neck. Underparts and patch around the eye white, female hornless although occasionally very rudimentary horns occur. Young males and females are yellowish fawn. Occasionally an old male remains in this color. Found on the open plains throughout India.

CHOUSINGHA OR FOUR-HORNED ANTELOPE. Reddish brown in color with white underparts. Distinctive as being the only antelope with two sets of horns although the front pair are sometimes merely a pair of stubs, not protruding through the skin. Female without horns. Found throughout most of India but prefer the hilly country where there is open forest suitable for grazing.

SAIGA *(Saiga tatarica)*

SAIGA. General color in summer buffy-yellow with white underparts. Whiter and with longer hair during the winter. Characterized by its remarkable inflated nose, short tail and amber-colored horns. Female hornless. Found from the country just north of the Caspian Sea west through northern Turkestan to Lake Balkish.

CHIRU, TIBETAN ANTELOPE. General color light rufous-brown. The face and the front surface of the limbs black or dark brown. The hair, especially in winter, soft and thick. Female without horns. Found on the Tibetan Plateau from Ladak to northern Tibet at altitudes ranging from 12,000 to 18,000 feet.

CHIRU *(Pantholops hodgsoni)*

17
THE NILGAI, THE KUDUS, AND THE NYALAS

NILGAI (*Boselaphus tragocamelus*)

The nilgai or blue bull is the largest antelope of India. The horns, in keeping with the antelopes of the bushbuck group, are found only on the male and are short and upright, curved toward the front. The nilgai prefers the plains or open ground with scattered trees. The animals generally go about in small parties but the old bulls are often solitary.

One of the most stately and beautiful of the antelopes is the greater kudu, also spelled koodoo. The long spiral horns have been recorded to over sixty inches in length and may make as many as three complete turns. It is an animal that prefers broken, hilly country and for this reason is very local in distribution. It is not a common animal in Kenya as it is found only in a few isolated hills, but in southern Ethiopia the writer had little trouble in finding it. From Tanzania south it occurs in fair numbers and in certain parts of Rhodesia it is considered a pest to native agriculturalists. It is one form of big game that has been able to withstand the incessant slaughter in South Africa. On the author's recent trip by motor north from Cape Town a kudu was the first big game animal observed from the road. A number of subspecies of the greater kudu have been described, differing slightly in color and amount of striping on the body.

As the name implies, the lesser kudu differs from the greater kudu in its much smaller size. Unlike its larger relative this kudu prefers to inhabit the hot, dry thornbush country of the lowlands.

The nyala, by many considered to be the most beautiful of the antelopes, is found in a very limited area. Similar in many respects to its relative the bushbuck, the nyala is larger, with more lyre-shaped horns, which have yellow tips. It is a forest animal seeking the deepest thickets during the day and coming out into the open only at dusk or early morning and is therefore seldom seen. When angry or excited, or when displaying before the female, the nyala will raise the long hairs along his back thus adding to his handsome appearance. Nyala appear to be most common in Zululand and southern Mozambique. A few are found in the Kruger National Park where their safety is assured.

The mountain nyala, also called mountain bushbuck and Queen of Sheba's antelope, is so like a greater kudu in appearance that it was not until 1910 that it was described. The natives call both animals "Agazan." The mountain nyala is found only on the Arusi Plateau of Ethiopia and is so local that few sportsmen have hunted it. The writer was very fortunate in having spent a month in the mountainous home of this animal, collecting a group for the American Museum of Natural History. Its scenic habitat and its stateliness, so in keeping with the environment, make the mountain nyala the author's favorite antelope.

The higher peaks of the Arusi Plateau rise to an altitude of 14,000 feet. Above the timber line these mountains are covered with a giant heather, which appears to be one of the main foods for the mountain nyala. On the sides of the mountains, directly below the heather, the forests consist chiefly of almost impregnable bamboo and it was in this cover that the animals rested. Early mornings and late afternoons they moved up onto the higher land to feed. Generally the bulls were alone, although at times family parties were observed.

Two or three cows were often seen together. They were remarkably sure-footed and when frightened would gallop over the rocky terrain with amazing speed. Their commonest neighbor was the little klipspringer.

NILGAI. Adult male, bluish or brownish gray, female and young reddish chestnut. A mane on the back of the neck in both sexes and a tuft of hair on throat only in the male. Found in Peninsular India, south into Mysore. Especially common in the Northwest and Central Provinces, eastern Punjab and Guzerat.

GREATER KUDU. Tawny in color with a number of white transverse bands along the body, and white chevron extending across face below eyes. Fringe of hair extending along back and on throat. Long spiral horns. Found locally from Cape of Good Hope north to Ethiopia on the east and Angola on the west.

GREATER KUDU *(Tragelaphus strepsiceros)*

LESSER KUDU *(Tragelaphus imberbis)*

LESSER KUDU. Darker in color than the greater kudu. White bands on sides and along back much more pronounced. White patch on the throat and one on the chest. No throat mane. Spiral horns less spreading. Found from central Tanzania north to southern Ethiopia and Somalia.

NYALA. Adult male dark brownish gray in color with narrow white transverse stripes along sides of body. Long fringes of dark brown hair from throat continuing down chest and along undersides of body. A mane of hair of the same color extends from the upper neck along the ridge of the back. From the shoulders to the tail the hairs are white-tipped. Lower legs tan color. Female and young male bright reddish chestnut. Found from Zululand in Natal north to Malawi.

NYALA *(Tragelaphus angasii)*

MOUNTAIN NYALA. Very similar in color and appearance to the greater kudu, including the mane along the back, but the hair on the throat is lacking. The horns of the adult male, unlike those of the kudu, make but a single spiral instead of two or three. Like the kudu the female is similar in color to the male. Found only on the mountains of the Arusi Plateau, south of Addis Abba, Ethiopia.

MOUNTAIN NYALA *(Tragelaphus buxtoni)*

18
THE BUSHBUCKS
AND THEIR RELATIVES

WEST AFRICAN BUSHBUCK
(Tragelaphus scriptus scriptus)

The bushbucks are small antelope standing under three feet at the shoulder. On account of their very extensive range, which includes most of Africa except the treeless plains and deserts, about twenty-four subspecies are now recognized. These differ chiefly in the color of the male, which may vary from a nearly black animal with very little trace of the white striping, the MENELIK BUSHBUCK, *T. s. meneliki* of the mountains of the Arusi Plateau of southern Ethiopia, to typical *scriptus* of West and Central Africa of a bright rufous color with the spots and stripes very conspicuous. The female is hornless and generally of a bright chestnut with distinct white stripes and spots. In the male a mane of long hairs extends along the back from the shoulders to the bushy tail. Bushbucks are animals of the forest and heavy bush and are great sulkers, coming out into the open to feed in the evening and early morning. They are generally found alone or in pairs and their distinctive bark is often the only clue to their presence. Although small they do not hesitate to stand their ground if wounded and can give a very good account of themselves.

The sitatunga is a member of the bushbuck group which has changed its mode of life and is exclusively a swamp and water animal. To facilitate travel over the soft muddy ground, the hoofs are greatly elongated and the lateral hoofs are more developed. The sitatunga spends the day hidden among the high reeds of the swamp, coming out into the open to feed only at dusk or at night. Hunting this animal is most difficult on account of its habitat and its ability to hide. It will frequently submerge, leaving only its eyes and nose above water.

The bongo is an animal of the thick forest. It is undoubtedly closely related to the eland. In common with this animal the female carries horns, and the oxlike tail is similar. However, it is a much more striking animal in coloration. The eastern representative, known as *B. e. isaaci,* is a denizen of the thick forests, preferring the heavy bamboo cover on the mountain sides. It is an exceedingly difficult animal to hunt and few are the sportsmen who have succeeded in securing one. For this reason, and on account of its beautiful coloring, it is often considered one of the most desirable trophies.

The eland is the largest of the antelope. Bulls often measure five feet eight inches at the shoulder. It is a plains animal but seems to prefer open forests when available, which afford it some protection. It associates in herds; at times large numbers congregate. Although of heavy build, its favorite pace is a long swinging trot which it can continue for miles. Its leaping ability is remarkable and it will often jump completely over another member of the herd. It is easily tamed and does well in captivity. When water is available, the eland drinks regularly; but when necessary, as with so many of the desert animals, it is able to secure enough water from the food it eats to keep it in perfect condition.

The giant or Derby eland goes about in small herds and is a browser, feeding principally on leaves. It has the habit of breaking branches with a twist of its horns to enable it more readily to reach this food. When disturbed this eland moves off on a swinging trot which the animal is capable of continuing for a considerable distance.

HIGHLAND BUSHBUCK *(Tragelaphus scriptus delamarei)*

WEST AFRICAN BUSHBUCK. The male of this typical form of bushbuck is a rich dark red in general color with black chest and underparts. White transverse stripes extending over the body, with a longitudinal line along the side. Numerous spots on face and body. The bushbuck is sometimes known as the harnessed antelope. Forests of West Africa from Senegal to Angola.

HIGHLAND BUSHBUCK. Slightly larger and browner than the West African bushbuck, with larger horns. Seal brown in color, white stripes and spots much subdued or lacking. Underparts and chest dark brown. Ranges from Tanzania north through Kenya to Lake Rudolf.

SITATUNGA. Male grayish brown in color and female generally reddish chestnut. White striping quite indistinct in many forms. Hoofs greatly elongated. Horns long and more openly spiralled than those of the bushbuck, the points with yellow tips. Found from Rhodesia north to the southern Sudan and in West Africa from the Congo River north to Senegal.

SITATUNGA *(Tragelaphus spekii spekii)*

BONGO *(Boocercus eurycerus)*

BONGO. Old male very dark chestnut, the head, neck and underparts almost black. Female and young male bright reddish chestnut. White transverse stripes along body. Horns spiralled and heavy with light tips, carried by both sexes. Ears large. Tail ox-like. Found in the forests of West Africa from the Congo to Sierra Leone and also in the highland forests of western Kenya.

COMMON ELAND (*Taurotragus oryx*)

COMMON ELAND. Uniform tawny in color but old bulls often look slaty gray because of the wearing off of the hair. A large dewlap hangs down from the lower neck. On the forehead a tuft of dark hair which differs in length according to the subspecies. White striping also occurs in some forms. Both sexes carry horns but those of the female are slimmer. In former times the eland was found from the Cape north to northern Kenya and on the west to Angola. At the present time it is not found in the Union of South Africa except on a reserve in Natal, in the Kruger National Park and on some farms in the Transvaal and the Orange Free State.

GIANT ELAND. Only slightly larger than the common eland but with longer horns and broader ears and the dewlap extending to the chin. Many bulls, especially of the typical race, have a black mane and long black hairs on the sides of the neck. The giant eland is found near the West African coast in Senegal, Gambia and Portuguese Guinea. It occurs also in north-eastern Congo, northern Central African Republic and the Sudan. The typical race or Derby eland is the western form. The Congo subspecies is *T. d. congolanus* and the Sudan race *T. d. gigas*. This last subspecies inhabits the Bahr-el-Ghazel region of the Sudan and is distinguished by rarely having the black on the neck of the males and in having the hair on the forehead brown instead of blackish.

GIANT ELAND (*Taurotragus derbianus derbianus*)

CHAMOIS (*Rupicapra rupicapra*)

19
THE GOAT-ANTELOPES

The chamois prefers the steep, rocky, forest-clad regions of the high mountains. It will often venture out into the open areas and snow fields but returns to the more protected areas during severe weather. It generally associates in small herds; and the older males, except in the breeding season, separate from the main herds and associate by themselves.

The Rocky Mountain goat is a denizen of the mountains above timber line. It prefers the steep mountainsides and cliffs, for it depends upon its wonderful ability to climb to protect it from its natural enemies. Unlike most of the larger animals of the high mountains, it does not descend into the sheltered valleys when winter sets in but remains aloft throughout the year, finding protection on the lea side of some cliff and seeking its food from the wind-swept mountainsides. The food consists of grass, lichens, mosses and the leaves and twigs of bushes. The heavy long hair and thick wool undercoat protect it from the wintry winds.

The Rocky Mountain goat generally associates in small herds. Unless the background is a snow-field, it is a very conspicuous animal and can be seen at a long distance. Generally one, but occasionally two, kids are born in April or May and when but a few days old are capable of following their mother.

143

The serow is a much larger animal than the goral, approaching the Rocky Mountain goat in size. The coarse black hair is characteristic. The animal has long, pointed ears and a distinct mane on the upper neck, which sometimes contains long white hairs. Unlike its American cousin, the Rocky Mountain goat, the serow prefers to live on the forest-clad slopes of the mountains. In Western Szechwan we found it in the thick bamboo that clothed the mountains, the home of the giant panda.

The famous Baie D'Along on the northeastern coast of Vietnam, once noted for harboring pirates, consists of numerous rocky islands, many of them dome-shaped and in size ranging from a few feet in height to more than a mile in length. These larger islands are well covered with brush. As we knew that serow inhabited some of these islands, we organized a hunt for the purpose of obtaining a specimen for the American Museum of Natural History. A number of native boys climbed to the summit of one of the larger of these islands and began rolling rocks down the sides. Soon a serow was disturbed and made its appearance and we were able to collect it while stationed on the shore. On another occasion, while camped in the mountains of western Szechwan, we were awakened by the sound of wild dogs running, and soon by the noise we knew that the animal they were trailing had been brought to bay. We hurried up the mountainside through the thick bamboo, and found where the dogs had just killed a large male serow which was soon added to our collection.

The goral is a much smaller animal than the Rocky Mountain goat. Over its very extensive range it differs slightly in size and color, and a number of forms have been described. The goral prefers rocky mountainsides covered with thick brush. Its coat so matches the surroundings that it is very difficult to observe. Two or three generally associate together,

144

although several may inhabit a certain terrain. The author's introduction to the goral was from a boat going through the famous Gorges of the Yangtze River, when four of the animals were observed among the bushes on the sheer cliffs above the river. Later in western Szechwan Province we spent a number of weeks in the country where goral were quite plentiful. Locating an animal was difficult. A movement would catch the eye. We would focus glasses on the exact spot but often it would be sometime before the animal itself could be located, so well did it match its environment.

CKY MOUNTAIN GOAT
(reamnos americanus)

CHAMOIS. General color tawny brown in summer, darker in winter, underparts and legs darker than sides, whitish buff patch on throat. Perpendicular horns hooking sharply to the rear. The true chamois and its subspecies are found in the Alps, Apennines, Carpathians to the Caucasus, Abruzzi Mountains, the Pyrenees and the Cantabrian Mountains of Spain.

ROCKY MOUNTAIN GOAT. White in color, at times with yellowish stain. In winter, hair long with an underwool. Beard on chin. Horns black and upright, curving toward the rear. Found from southern Alaska south through British Columbia and western Alberta to Washington, Idaho and western Montana.

SEROW. Hair long and coarse. General color black with a grizzled appearance. Color varies with the different forms. Horns comparatively short, curving backward. Found from Sumatra and the Malay Peninsula north through Vietnam, Thailand, Burma, Assam, the eastern Himalayas and China.

SEROW *(Capricornis sumatraensis)*

JAPANESE SEROW (*Capricornis crispus*)

JAPANESE SEROW. A small serow with thick, woolly coat and hairy tail, approaching the goral in appearance. General color blackish gray with lighter underparts. Limited to Japan with a close relative, the FORMOSAN SEROW, *C. swinhoei*, occurring on Formosa.

GORAL. Color brownish gray, generally with a throat patch of either white or yellow. Legs generally darker. Horns small and curving toward the rear. Found from the Himalayan Mountains east through the mountains of West China and north to Korea.

GORAL (*Noemorhedus goral*)

MUSKOX *(Ovibos moschatus moschatus)*

20
THE MUSKOX
AND THE TAKIN

The muskox is a creature of the Arctic tundras of North America. Three forms are recognized, the most southern of which is the BARREN GROUND MUSKOX, *O. m. moschatus*. It ranges from the northwestern shores of Hudson Bay north to Carnation Gulf. The HUDSON BAY MUSKOX, *O. m. niphoecus*, is slightly darker, especially about the head in the adult male, while the female and young have some white on the face. Its range is north of the Barren Ground muskox from northern Hudson Bay to the Arctic Coast. The WHITE-FRONTED MUSK-OX, *O. m. wardi*, occurs on the islands of the Arctic Sea from Banks Land and Grant Land eastward to the northwestern and northern coasts of Greenland. It is slightly lighter in color and has more white about the head.

149

At different times the muskox has been thought to be related to the sheep, to the Rocky Mountain goat, and more recently to the bison. It generally keeps together in small herds, chiefly for protection. Besides man, its chief enemy is the wolf; and at the first sign of danger the herd form a circle with the young animals in the center, while the old ones face outward, thus presenting a circle of horns to the enemy. Although this was effective against the wolf, it was useless against rifles; and the herd could be slaughtered without difficulty. At the present time the Barren Ground muskox has been greatly reduced in numbers and has disappeared from much of its former range. The white-fronted muskox lives in less accessible districts and is today the most common form.

Like those of the caribou, the hoofs of the muskox are broad, thus helping the animal to traverse the snow. The muskox has supplied meat to numerous explorers, and many a man owes his life to this animal.

The muskox became extinct in Alaska before 1865. In 1930 a herd of animals was captured in Greenland. From there they were shipped to Oslo, Norway, to New York, across country to Seattle, and then north to Seward, Alaska. They withstood this trying trip remarkably well and were transferred to Nau-nivak Island where they had some two thousand square miles in which to roam free from natural enemies. By 1943 the animals had increased to well over one hundred individuals. It is hoped that with this herd as a nucleus enough of these animals may be transferred to the Arctic barrens of northern Alaska to repopulate the land previously occupied by the former herds.

The long hair and thick underwool afford the muskox good protection from the cold. Unlike the caribou, the muskox is not migratory. Its chief food is grass, but lacking this, it feeds on mosses and lichens and browses on willows. A single calf is born in late May or early June.

The takin, which up to a few years ago was thought to be related to the Rocky Mountain goat, goral and serow, is now believed to be nearer to the muskox in classification. It is an animal of the mountains but prefers thick cover, such as rhododendron and bamboo thickets. In Szechwan, where we spent some time following its trails, we found that it wandered over the mountainside often above timber line, but to feed preferred to descend to the valleys which contained small swamps. In these valleys a certain broad-leaved plant was abundant. This evidently was a favorite food, for wherever this plant was found, one was pretty sure to discover numerous tracks. Its food also consists of grass, bamboo shoots, and leaves and twigs of certain bushes. The sides of the mountains were often made up of loose, broken stones and boulders, and it was remarkable with what ease and speed such a short-legged, heavy and seemingly awkward beast could cover the difficult terrain.

The takin live in small herds, and when travelling carry their heads low, their large convex nose, heavy, rounded body and short legs giving them an odd appearance.

The typical or Mishmi takin is found in the mountains of Bhutan, northern Assam and northern Burma. The Szechwan form is found in western Szechwan and neighboring eastern Tibet. The Shensi takin inhabits the highlands of southern Shensi, north into Kansu Province, China.

THE MUSKOX. Nearly black in color except for a light saddle mark, and brownish black on the back. Long hair reaching almost to the ground, and a heavy underwool. Horns, with a broad frontlet that nearly meets, curve downward close to the head, then outward and upward. Horns of the female similar but smaller. The muskox is found only in Arctic North America from the western shore of Hudson Bay northwest to about the one hundred and twentieth meridian and north into Grant Land and Banks Land and the eastern and northern coasts of Greenland.

THE TAKIN. A short-legged, heavy-built animal. In color varying according to the subspecies. The Szechwan race, *B. t. tibetana,* is a richly colored animal, the head, neck and shoulders being a rich golden yellow, and the hinder part of the body and legs grayish and black. Nose black. Female of a more silvery tone. The Mishmi or typical form, *B. t. taxicolor,* is darker, while the SHENSI TAKIN, *B. t. bedfordi,* is much lighter and lacks the black nose and other markings. The base of the horns arises from the forehead; the horns turn outward and then to the rear and slightly upward, the points turning slightly inward. Horns of the female are similar but smaller.

TAKIN (*Budorcas taxicolor*)

IBEX *(Capra ibex)*

21
THE WILD GOATS

SPANISH IBEX *(Capra pyrenaica)*

PERSIAN IBEX *(Capra hircus aegagrus)*

IBEX, STEINBOCK. General color brownish gray with chin, upper throat and underparts darker. Lower legs dark brown. Hair longer on back of neck, forming a small mane on the older bucks. Beard of male short and small and confined to chin. Horns long, sweeping backward in long curve, front surface broad with the transverse protuberances well marked. Formerly found throughout the Alps of Switzerland, northern Italy and neighboring highlands. Now restricted to a few small protected herds on the Italian side of Monte Rosa. Recently reintroduced into several reserves in Switzerland.

SPANISH IBEX, SPANISH TUR. General color grayish brown with a dark dorsal line. Outer sides of limbs, shoulders and chest black or darker brown. A darker line along lower body. Underparts white. Old male much darker than female and young male. Beard black, long in winter but much reduced in summer. Horns unlike the ibex type and form an open spiral rising in an upward direction, then outwards and backwards with the tips slightly upward, out over the neck. Triangular in circumference with sharp inner edge, quite similar to the Caucasian tur. Found in the Pyrenees and mountainous districts of Spain and Portugal.

PERSIAN IBEX, PASANG. Brownish gray in winter and reddish brown in summer with white underparts. Dorsal stripe, collar, chest and forepart of legs darker brown. Underparts and rear of legs white. Horns scimitar-shaped with sharp inner front edge and several widely-separated knobs along the rear half. Beard long and restricted to chin. Generally believed to be the ancestor of the domestic goat. Several subspecies range in the mountains of southwestern Asia from the Caucasus through Asia Minor and Iran to Baluchistan and Sind.

155

ASIATIC IBEX *(Capra ibex sibirica)*

ASIATIC IBEX. General color grayish brown with a dark dorsal stripe. Lighter in winter. Beard of male long and pointed and confined to the chin. Horns long with front surface broad with numerous transverse knobs. A number of subspecies occur which range throughout the mountains of central Asia from the Tien Shan and the Altai to the Himalayas in Kashmir.

HIMALAYAN IBEX. General color pale brownish white with a pale brown dorsal stripe. The horns average a wider spread than the horns of the typical Asiatic ibex. Himalayan Mountains from Kashmir east through Nepal.

156

HIMALAYAN IBEX *(Capra ibex sakeen)*

ABYSSINIAN IBEX (*Capra walie*)

ABYSSINIAN IBEX, WALI. One of the most striking of the ibex. Upper parts chestnut-brown, grayish white below. Body hairs have dark bases producing a grizzled appearance. Short beard brownish black, as are legs. It differs from all the other ibex in having a bony swelling of the forehead. Horns large, sweeping and heavy at the base, the transverse knobs distinct. Found only in the mountains of the Simien Plateau, Ethiopia.

NUBIAN IBEX. Upper parts brown. Line along back, beard, chest, outersides and front of legs brownish black. Innersides of legs and band above each hoof, white. Beard long and full. Horns long and slender. Front surface narrow with numerous well-formed transverse ridges. Found in the Red Sea Hills and the coastal mountains of North Africa. Subspecies found in Arabia and Palestine.

NUBIAN IBEX (*Capra ibex nubiana*)

ASTOR MARKHOR *(Capra falconeri falconeri)*

ASTOR MARKHOR. General color in winter gray, in summer rich reddish brown but the old male is whitish. In winter, hair long and silky with little underfur. A heavy beard, black in front, gray-brown behind, extends from the throat to the chest. The large horns of this subspecies form an extremely open spiral with relatively few turns. Found in the mountains of Kashmir at Astor and Balistan.

CABUL MARKHOR. In color similar to the Astor markhor but slightly smaller. Horns are nearly as straight as those of the SULEMAN MARKHOR, *C. f. jerdoni,* in which form the horns make a number of complete turns and are perfectly V-shaped. The horns of the Cabul markhor show a very slight opening spiral. Found in the Cabul district of Afghanistan.

158

CABUL MARKHOR *(Capra falconeri megaceros)*

CAUCASIAN TUR *(Capra caucasica)*

CAUCASIAN TUR. In winter, general color dull brown. Chin, tail-tip, front and innersides of hind legs and front of forelegs below knees, blackish brown. Beard confined to chin, short and broad, curling forward. Similar in color to body. Horns massive, widely separated at base, with rounded front surface, first going in an outward and slightly upward direction, then backward, downward, inward, and the points turning upward. Found in the Caucasus.

TAHR, HIMALAYAN TAHR. Dark reddish brown with hair on the chest, neck and shoulders of the male very long, reaching to the knees. Hair in the fore part of this ruff paler. Head and legs dark brown. Horns short like those of the Nilgiri tahr but differing greatly in shape. They almost touch at their base but are greatly compressed and flattened for a short distance with the sides ridged transversely. Horns are sharply ridged, curve back sharply and are slightly spread. Found south of the main range of the Himalayas from Bhutan to Kashmir.

TAHR *(Hemitragus jemlahicus)*

NILGIRI TAHR *(Hemitragus hylocrius)*

NILGIRI TAHR. General color dark brown with dark dorsal stripe and paler underparts. Large saddle-shaped patch on the back, grizzled white, almost pure white in old male. Hair short and thick, forming a short mane on the neck and shoulders of the male. Horns short, about equaling the length of the head, those of the female only slightly shorter. They nearly touch at their bases, rise almost parallel, then turn backward, spreading slightly. The inner surface is flat and the outer convex. Found in the mountains of southern India.

ARABIAN TAHR. The smallest of the tahr, about twenty-four inches at the shoulder. The grayish-brown hair is relatively short. Horns more slender than those of the other tahr. Inhabits the mountains of southeastern Arabia.

ARABIAN TAHR *(Hemitragus jayakari)*

MOUFLON (*Ovis musimon*)

22
THE WILD SHEEP

RED SHEEP (*Ovis orientalis*)

MOUFLON. General color reddish brown with a light saddle-patch in the male. Underparts white. Horns of the adult ram generally form about one complete turn with tips bending forward and outward. Female infrequently has horns. A small sheep about twenty-seven inches at the shoulder. Found on the islands of Corsica and Sardinia.

RED SHEEP, ASIATIC MOUFLON. Small sheep of reddish color with white underparts. White saddle-mark in the older males. Horns large for so small a sheep, sweeping upward and outward, then downward and inward so that the tips are over the shoulders. Female hornless. The red sheep and its subspecies are found in the mountains of Cyprus, Asia Minor, Armenia, Transcaucasus and Iran.

ROCKY MOUNTAIN SHEEP, BIGHORN. Grayish brown on upperparts, darker on shoulders, neck and legs. Nose light. Dark dorsal line to tail. Underparts and rump patch lighter. Lighter in spring and summer. Horns massive, forming at least a complete turn in the older ram, but without much spread. Tips of horns generally broken or broomed. Length of over forty-nine inches has been recorded. A number of subspecies are known. Mountains of western North America, from central British Columbia and Alberta south into Mexico and the peninsula of Lower California. Forms of the bighorn sheep are found in Kamchatka and parts of Siberia.

ROCKY MOUNTAIN SHEEP (*Ovis canadensis*)

WHITE SHEEP *(Ovis dalli)*

WHITE SHEEP, DALL'S SHEEP. Smaller in size than the bighorn. Typical Dall's sheep white throughout or slightly washed with dusky. Farther south in its range the pelage becomes darker and there is a gradual intergradation to the STONE'S SHEEP, *O. d. stonei,* a very dark gray sheep of northern British Columbia. The horns of the Dall's sheep are more slender than those of the bighorn and have a tendency to spread, while the tips are less likely to be broken or broomed. The species ranges from the mountains of northern Alaska south into northern British Columbia. The true white sheep is found in the mountains of Alaska and as far south as the Alaskan-Yukon border where the intergradation begins to take place.

SIBERIAN ARGALI. Largest of the wild sheep, with heavy, massive horns. Winter coat light brown with grayish wash. Face and underparts gray. Distinct white tail disk. In summer, coat is of a lighter color. Horns often forming more than a complete circle and curving outward at the tip. The record length of the front curve of the horn is sixty-two and one-half inches and the circumference of another horn is twenty-one and one-half inches. A third set of horns has a forty-two and one-half inch spread. Formerly found from the Baikal Mountains of southern Siberia west to the Altai. It is now rare in the eastern part of its range although it still occurs in some numbers in the Altai.

SIBERIAN ARGALI *(Ovis ammon ammon)*

ALATAU ARGALI *(Ovis ammon karelini)*

ALATAU ARGALI. In the general region of the Tien Shan Mountains a number of forms of sheep have been described, many of them intermediate in appearance between true *ammon* of the north and *polii* of the south. The Alatau argali is one of these, the typical species being restricted to the Alatau Mountains. This sheep has a small rump patch and the tips of the horns diverge from the sides of the head. The largest recorded horn is fifty-three inches long, with the circumference at the base sixteen and one-half inches and a thirty-three and one-quarter inch spread for the set.

MARCO POLO'S ARGALI. Noted for the extreme length of its horns which have been recorded to seventy-five inches. A light-colored sheep, the face, throat, underparts, flanks and legs white, the upper body washed with brown. In winter, hair is longer and darker on the body and there is a dorsal stripe, the longer hair of the chest and throat forming a white ruff. This is a large sheep, only slightly smaller than the Siberian argali. The horns are rather slender for their extreme length, frequently making over one and one-half circles, the points extending outward. A spread of fifty-six inches has been recorded. Found only on the Pamir Plateau.

MARCO POLO'S ARGALI *(Ovis ammon polii)*

SHAPO *(Ovis orientalis vignei)*

PUNJAB URIAL *(Ovis orientalis punjabiensis)*

SHAPO, LADAK URIAL. Upper parts vary from rufous-brown to gray in summer, lighter in winter. Tail, underparts and limbs whitish. Ruff of the throat and chest black, some of the hairs sometimes white. Horns of a spreading type, rising close together on the head, growing in a circular form at first toward the rear and outward, then turning in and forward, the tips pointing back toward the face. Found in the mountains near the India-Tibet border, in the vicinity of Astor and Ladak.

PUNJAB URIAL. Slightly smaller than the Ladak urial and of a redder color with well-developed white ruff. Horns only slightly spiral, forming a more compact circle, the tips approaching the vicinity of the eyes. Found in the Punjab and Sind and west into Afghanistan.

AFGHAN URIAL. Upperparts yellowish brown, underparts and inside of limbs white, face bluish gray. Beard black with the long hair from the jaw to the chest interspersed with white and gray. Slightly darker in winter. Horns forming a more open spiral than the shapo and tending to turn outward at the tips. Found in Afghanistan, Baluchistan, and neighboring territory.

AFGHAN URIAL *(Ovis orientalis cycloceros)*

BHARAL *(Pseudois nayaur)*

BHARAL, BURREL, BLUE SHEEP. General color blue-gray, underparts and inside of limbs white. Stripe down the face, chest, fronts of both the fore- and hind-limbs, except the knees, and stripe along the lower sides, separating the dark upper body color from the white of the underparts, black. Horns very distinctive, being the same general type as those of the Caucasian tur. They are quite round with a general outward sweep, first extending slightly upward and outward, then downward and backward and finally the tips turning inward. The bharal has characteristics of both the sheep and the goats, and affords a connecting link between the two groups. The highlands of central Asia from northern India to the mountains of western China.

AOUDAD, ARUI, BARBARY SHEEP. Color tawny, male with a fringe of long hairs extending from the throat to the knees. Tail long and hairy. Horns of much the same shape as those of the bharal but having a distinctive keel on the front surface near the base. The only wild sheep found in Africa. The mountains of North Africa from Morocco to Egypt, and northern Sudan to Tchad and Mali.

AOUDAD *(Ammotragus lervia)*

23
THE BUFFALOES

CAPE BUFFALO (*Syncerus caffer*)

The Cape buffalo is never found far from water. It is a grazer but will adapt itself to the most varying conditions so long as water is available. It is equally at home in the mountainous forests or in the semi-desert country of the northern Kalahari. While camped in this latter country at the edge of a small pan or water hole, only about a hundred yards in extent and about three feet deep—the only water supply for miles around—we used to be entertained by nightly visits of buffalo and elephants. Buffalo are considered by many sportsmen to be one of the most dangerous game to hunt. Occasionally one will charge without provocation, but as a rule they will beat a hasty retreat. However, following up a wounded animal in thick cover is risky business and should not be undertaken by a novice.

Buffalo go about in herds, sometimes in extensive numbers. Old bulls may lead a solitary life, or a number of them may join up into a small herd.

Near the close of the last century an epidemic of rinderpest broke out among the buffalo herds in East Africa. Thousands of the animals died, but the few survivors soon increased until, at the present time, buffalo are again common over much of the infested territory.

The Chad buffalo is now considered a subspecies of the

small forest buffalo but it shows a general approach to the larger eastern animal. The Chad buffalo is intermediate in size, color and type of horns between the Cape buffalo toward the east and the forest buffalo on the west. Although specimens may be very dark brown, they are never black as is the Cape buffalo. The horns are much heavier and not upright as in the case of the forest buffalo; they are more spreading as in the Cape buffalo. However, the horns are shorter and do not have the downward curve.

The red or forest buffalo of West Africa, as the name implies, prefers the forested areas although it is sometimes found out in the open country. A number of subspecies have been described, differing in size, and the color may vary from the yellowish red of the typical form to brown. In one subspecies the cows are of the typical color but the bulls are dark.

The Indian buffalo is best known as a domesticated animal. It has been introduced as a draft animal throughout most of the tropical countries. In many of these places the buffalo has resorted to a semi-wild condition and it is difficult to ascertain, in the countries where the wild buffalo still occur, which are the truly indigenous and which are the feral. Certain strains of the domesticated buffalo have the drooping horns, somewhat of the type of the Cape buffalo.

In Borneo a small buffalo, *B. b. hosei,* occurs, which is supposedly wild and indigenous to that island.

The tamarou prefers marshes and districts of heavy cane but is also found in the forests. It is very difficult to hunt on account of the thick cover of its habitat, and it has the reputation of being exceedingly aggressive.

The anoa is a forest animal preferring mountainous districts. Very little has been published concerning its habits. It is more solitary and seldom found in herds, preferring to go about singly or in pairs.

170

CHAD BUFFALO *(Syncerus nanus brachyceros)*

CAPE BUFFALO. General color of hair black throughout, although often sparse, exposing the dark brown skin. Young calf a dark reddish color. Horns massive, covering the crown of the head with a horny plate. Horns extend outward and downward, then upward and inward. A spread of fifty-six and one-half inches has been recorded. The Cape buffalo formerly was common throughout southern and eastern Africa. In South Africa, south of the Zambesi, buffalo have become rare; but in Zambia, Tanzania, Kenya, and the Sudan they are still numerous in suitable localities.

CHAD BUFFALO. Color gray or brown. Horns heavy but rather short and without the downward sweep of true Cape buffalo. Found in the vicinity of Lake Chad, West Africa.

RED OR DWARF BUFFALO, BUSHCOW. General color yellowish red. Lower legs black. Horns much smaller and less massive than those of the Cape buffalo. They are broad at their base and grow in an outward, backward and upward direction, the tips turning in. Confined to West Africa from Angola north.

RED BUFFALO *(Syncerus nanus nanus)*

INDIAN BUFFALO *(Bubalus bubalis)*

INDIAN BUFFALO. Color blackish gray, sometimes with brownish tinge, hair often very sparse. Horns long with many irregular transverse grooves. Bases start at the sides of the forehead extending outward and backward and then tip inward in a wide sweep. India and neighboring countries.

TAMAROU. General color dark gray. White markings similar to those of the anoa. Horns short but broader than those of the anoa with many deep, transverse grooves. Stands about six inches higher at the shoulder than the anoa. Found only on the island of Mindoro, the Philippines.

TAMAROU *(Anoa mindorensis)*

172

ANOA. The smallest of the buffalo, measuring about three feet three inches at the shoulder. Hair of adults dark brown or black, often very sparse. Young animals thickly covered with brown hair. White crescent on the throat and white spots on the lower jaw. Short horns triangular at the base, following the plane of the face in direction, spreading slightly. Confined to the Celebes.

ANOA *(Anoa depressicornis)*

24
THE BISON AND THE YAK

The American bison or buffalo is an outstanding animal. A large bull stands as much as six feet at the highest point of his shoulders. The whole forepart is massive with long woolly hair extending over his face, head, shoulders and the upper forelegs. The long hairs of his chin form a long beard. However, the hips and hind quarters are proportionately small and devoid of the long hair, thus forming a distinct slope from the hump to the tail.

The bison at one time was found as far east as the Allegheny Mountains. This EASTERN BISON, *B. b. pennsylvanicus,* spent the summers as far north as western New York State and Lake Erie. The range extended south through western Pennsylvania as far as northern Georgia. There were seasonal migrations north and south. Most of these bison disappeared

175

AMERICAN BISON *(Bison bison)*

after the Revolutionary War when this country was opened up. The last bison in Pennsylvania was believed killed in 1801. Bison roamed the western plains in countless thousands, but in the early eighteen hundreds the West was being opened up and by 1830 the slaughter had begun. Bison were killed for their hides and tongues. When the transcontinental railroad pushed its way across the plains it spelled the doom of the bison. By 1900 there were less than one thousand living bison and it was feared that they would soon become extinct. Through the efforts of the American Bison Society and other interested persons the bison has made a remarkable recovery. They breed readily on reserves and even in small quarters so that with protection and under semi-domestication the herds were built up. Today there are very few truly wild bison in existence but there are numbers on reservations in a semi-wild state.

The WOOD BISON, *B. b. athabasca,* is a large dark form that is found in the region south of Great Slave Lake. This animal is still found in the wild state and, as the name implies, inhabits the more open forests and muskeg swamps. In the southern part of the wood bison's range some domesticated plain bison have been introduced so that it is doubtful whether the wood bison will long remain in a pure state.

The European bison, referred to as the wisent by the Germans, is sometimes called the auroch. This is misleading, for the true AUROCH, *Bos taurus,* was the wild ox of Europe, an animal of somewhat the general appearance of certain breeds of the Spanish fighting bulls but much larger. It is believed to be the ancestor of most of the domestic breeds of cattle. It evidently became extinct about the middle of the sixteenth century.

The European bison is a more rangy animal than the American bison and lacks the massive head and shoulders, while the hind quarters are heavier. It is chiefly an animal of the forest and feeds more commonly on foliage, bark and twigs, but it frequently grazes like its American cousin. Formerly this bison was found throughout Europe, but in modern times it has been restricted to two districts, the Bielowitza Forest in Lithuania and the Caucasus. The CAUCASIAN BISON, *B. b. caucasicus,* was considered a subspecies on account of its slightly smaller size and other minor differences. It became extinct by 1928. The Lithuanian bison was more fortunate but its numbers have been greatly reduced. Much labor and time have been spent in trying to breed up the last remaining animals, with some success, but revolutions and wars have taken their toll. In January, 1947, it was reported that there were slightly over one hundred pure-bred European bison left, the majority in Poland.

The yak originally had an extensive range through the Tibetan Plateau from Kashmire to western China and large herds were reported. In Kashmir it was afforded protection; but over most of Tibet and in Kansu Province, China, it has been so harassed by native hunters, who hunt it for both meat and hide, that it is now restricted to the more remote and desolate regions. As is the habit of so many of the wild cattle, the female yaks with their young and a few master bulls make up the large herds, with most of the old bulls either leading solitary lives or else joining up with a few other bulls in small groups.

In Tibet the yak has for years been successfully domesticated and has been introduced into Turkestan and neighboring regions. The domestic yak is smaller than the wild one and is often without horns. It is used as a pack animal and also for riding. Its flesh and milk are used for food; the hair and wool for cloth. It is as important to the natives of this bleak country as the camel is to the natives of the deserts.

EUROPEAN BISON *(Bison bonasus)*

AMERICAN BISON. Brown, the long hair of the head and shoulders being darker. High at the shoulders, hind quarters small in comparison. Formerly found from Pennsylvania west to the Rocky Mountains and from Great Slave Lake on the north to Mexico on the south. Now limited to parks and reserves in the plains regions of the United States and Canada.

EUROPEAN BISON, WISENT. Well-clothed with long chestnut-brown hair, especially about the head and shoulders. Generally a distinct mane. Horns go outward, upward, forward and points turn inward. They are more slender than those of the American bison and have been recorded up to twenty inches long. Formerly with quite an extensive range in central Europe. For many years restricted to the Caucasus and Lithuania. Now, however, reduced to a few captive individuals.

YAK. Uniform blackish brown in color with a fringe of long hair covering chest, shoulders, sides, flanks and tail. Domesticated individuals often have much white in their coats. Horns long, extending outward and upward, then forward, inward and upward. A horn thirty-eight and one-fourth inches has been recorded. Originally found from Kashmir, across the Tibetan Plateau to Kansu Province of China.

YAK *(Bos grunniens)*

25
THE WILD CATTLE

GAUR *(Bos gaurus)*

The gaur prefers the mountain forests; but in Indochina, where I had the opportunity of observing this animal, they came down from the hills shortly after the natives had burned off the grass and canes of the lowlands, and here they could be observed in the late evening and at night feeding on the newly-sprouted grass. If disturbed, they took refuge in the eight-foot stands of elephant grass from where it was impossible to dislodge them without the aid of elephants. Gaur generally go about in small herds of ten or twelve animals but often larger herds are observed. These herds generally con-

tain one or two bulls, but most of the old bulls lead a solitary existence or may join up with one or two other males. The Malayan form of the gaur is known as the Sladang.

The gayal is a domesticated form of the gaur, which is kept by the hill tribes along the border of Assam and Upper Burma. These animals are allowed full liberty and roam the neighboring hills throughout the day returning at night to the villages.

The kouprey is of special interest because it remained unknown to science until it was described from a captive individual in the Vincennes Zoological Gardens in France in 1937. Previous to that time, reports had come in about a strange wild forest ox which occurred in Cambodia, but this animal was believed to be some sort of hybrid. In later years a number of these animals have been taken or observed and there is still doubt as to their exact status. The kouprey associates with the herds of wild banting, which fact leads one to suspect that if the animal is a hybrid, it is evident that one of the parents is a banting. Gaur and banting associate in the same country, and according to Francois Edmond-Blanc, who collected the first specimen of the kouprey to reach a museum for scientific study and who has studied the animals in the wild, the banting and gaur have been found associating in the same herd. The hoofs and hump of the kouprey are gaur-like but the great dewlap is lacking in both of these animals. In the dewlap and shape of head the kouprey resembles some of the local domestic cattle. Again in habits, such as its wallowing in mudholes, this animal resembles the buffalo; but both the banting and the buffalo lack the distinctive hump. Mr. Edmond-Blanc believes, however, that the kouprey is a hybrid, but he will not commit himself as to the exact cross. Other scientists believe the animal to be a perfectly true species and there are numerous sound reasons on which they can base their arguments. A thorough study of the life of this animal will be necessary before any definite conclusions can be made. The kouprey is undoubtedly a very rare animal and should have absolute protection; otherwise it may become extinct before its true status becomes known and one of the latest of nature's riddles has been solved.

Like the gaur the banting travels about in small herds, generally led by an old bull. However, it prefers the lower, flatter land to the rugged mountains and often resorts to the extensive bamboo forests and thickets. From our experience it seems less wary than the gaur. While motoring one evening in Cambodia, we stopped our car to watch a herd of these animals cross the road in front of us.

GAYAL (*Bos frontalis*)

THE GAUR, INDIAN BISON. The largest of the wild cattle, the bulls often exceeding six feet at the shoulder. General color brownish black with legs white below knees and hocks. Forehead gray. A prominent ridge along the shoulder. Horns broad at the base and heavy, of a greenish color with black tips, and may be over thirty inches in length. Found from India east into Assam, Burma, Thailand, Cambodia, Laos, Vietnam, and south into the Malay Peninsula.

THE GAYAL, MITHAN. Slightly smaller than the gaur, dorsal ridge less elevated and dewlap more developed. The gayal is the same general color as the gaur. Horns differ in lacking the greenish tone, and extend outward and upward, not curving inward at the tip. Found in the hilly country of eastern and southern Assam and Upper Burma.

THE KOUPREY OR FOREST OX. Slightly smaller than the gaur. General color dark brownish black, grayish on the shoulders and hips. Female and young gray. White below knees and hocks. A long dewlap extending from throat and chest. Horn curve is first backward and outward then sweeps upward and forward, and the tips turn inward and slightly backward. An interesting characteristic of the horns of the bull is the frill of shredding horn a few inches below the tips. Known only from central Cambodia.

KOUPREY (Bos sauveli)

...ESE BANTING *(Bos banteng birmanicus)*

BURMESE BANTING, TSAINE. Slightly smaller than the Javan banting, the bulls being of a chestnut color but slightly darker than the cows. Face gray. Occasionally an old bull will take on the dark color of the typical race. Found in Burma, Thailand, Cambodia, Laos, and Vietnam.

185

JAVAN BANTING, BANTIN. Old bull blackish brown but differs from the gaur in having a large white rump patch. Lower legs white. Young bull and cow bright reddish brown. Height at the shoulder up to five and one-half feet. The banting is of a lighter, more cow-like build than the gaur. Horns of the bull more rounded and less flattened at the base than the gaur's and distinguished in having a horny shield extending along the crown of the head between the bases of the horns. Found in Java, Borneo, Sumatra, Bali, and formerly in the Malay Peninsula, but it is believed that the true banting no longer exists in the latter place. The banting readily interbreeds with the native cattle and the animals now inhabiting that country contain domestic blood.

JAVAN BANTING (*Bos banteng banteng*)

26
THE HIPPOPOTAMUSES

PYGMY HIPPOPOTAMUS (*Chœropsis liberiensis*)

The pygmy hippopotamus is pig-like in appearance and stands about three feet in height. It is more solitary in habit than its larger cousin, living either alone or in pairs. It inhabits the deep forests and frequents the banks of the smaller rivers where it spends much of the day concealed in the vegetation along the shore or in hollows under the banks. It is chiefly nocturnal, leaving the water to feed in the nearby forest during the night. It is also found in swampy forest land and is less aquatic than its big cousin. On account of its nocturnal habits and its limited and inaccessible range, it has seldom been seen

by white men. In 1912, Hans Schomburgk, an agent for Hagenback, the noted animal dealers of Hamburg, Germany, succeeded in capturing three individuals. A pair of these was sold to the New York Zoological Society. Fortunately they stood captivity well and a number of young were raised.

The hippopotamus is distantly related to the pigs, but differs in its large head with a broad nose, the four sub-equal toes all of which help bear the weight of the animal, and in tooth characteristics. Like the pigs, the canines form large tusks but these do not extend outside the lips. The hippopotamus is adapted to its aquatic life. The nostrils are rather high on the nose and elongated and can be closed at will to keep out the water. Like the nostrils, the small eyes are slightly raised and are on the upper surface of the head, making it possible for the animal to see and breathe while exposing little more than its nostrils and eyes. A big bull will weigh three and one half to four tons.

The hippopotamus is still common in many parts of Central Africa. While camped on the shores of Lake Abaya in southern Ethiopia, we had a fine opportunity to observe a herd of about thirty individuals. Throughout the day they spent their time out in the lake or on some secluded sand bar. There were a number of calves in the herd and the youngest ones were often seen resting on their mothers' backs. Our camp was pitched under an escarpment in front of which was a gentle slope to the water's edge, covered with short grass. It was evident that the hippos were accustomed to come ashore nightly to feed on this grass. After the first day or two, our camp did not seem to bother them, as at dusk they would come ashore and feed with our mules. Behind our camp, the escarpment rose to about two hundred feet, steep enough so that the mules had to pick their way. Much to our surprise we found that the hippos had crossed over this into the little valley on the other side.

On the west shore of the lake, thick thornbush grew down to the water's edge. Hippo trails penetrated this thornbush in every direction; in fact the only possible way to go through was to follow the hippo trails. Other game animals, such as water buck, also used these trails. One morning while crawling through a particularly dense thicket on my hands and knees, I came face to face with a leopard.

As a rule hippos are not dangerous and a herd may live in the river near a native village in perfect equability. This is not always the case, for at times a hippo may turn sportive and overturn a native canoe. Such an instance occurred while I was in Rhodesia. A bull hippo had taken up his abode at a section of the river where natives were constantly crossing to the market. After a number of unfortunate adventures, orders were given to destroy the animal. This was welcome news to the natives in more ways than one, for they were very fond of its flesh. In some of the larger lakes the natives hunt the hippo from canoes. This is an exciting and dangerous sport. A number of canoes will gather together, each containing paddlers and men armed with spears to which are attached long lines tied to inflated skins. A spear is thrown into an animal and the float shows the natives its whereabouts below the surface of the water. When the animal comes up to breathe, more spears are hurled until finally the creature succumbs. At times the animal may become enraged, overturn a boat and attack the natives in the water.

If much molested, hippos become shy, hiding away during the day, coming out to feed only at night. Their chief food is aquatic vegetation, grass, reeds, etc., but at times they will visit the native gardens.

PIGMY HIPPOPOTAMUS. Much smaller than the large hippopotamus. General color of its nearly hairless skin is a shining grayish blue-black. Head smaller in proportion to the body than in the large hippopotamus, the eyes and nostrils not raised above the plane of the head to the great extent so characteristic of the larger animal. It further differs in generally having only one instead of two pairs of lower incisors. Toes slightly more separated and spreading. Found in Liberia, the Ivory Coast and Sierra Leone.

HIPPOPOTAMUS. Appears to be almost hairless, although short hairs can be felt over much of the body if brushed by the hand. Tufts of hair on the upper lip, the tips of its ears and the end of its tail. General color grayish black with underparts and folds in the skin flesh-color. Originally found in the large rivers and lakes from the delta of the Nile south to the Cape of Good Hope. Now absent from the Nile north of Khartoun, and extinct south of the Orange River in South Africa. However, in certain sections of Central and East Africa it is still abundant.

HIPPOPOTAMUS (*Hippopotamus amphibius*)

EUROPEAN WILD BOAR *(Sus scrofa scrofa)*

27
THE WILD BOARS
AND THE ASIATIC WILD PIGS

The wild boar has a very extensive range throughout much of Europe, Asia and northern Africa, and a number of subspecies have been described. It is the ancestor of most of our domestic pigs. It prefers forests and obtains much of its food by rooting in the ground. In many parts of the world it causes much damage to cultivated crops, not only in the amount it eats but also by its habit of rooting, which destroys much more than is consumed. The wild boar goes about in small parties called "sounders." It is pugnacious and well able to give a good account of itself if necessary, doing great damage with its sharp tusks. The young, which vary in number from six to ten, are brown with black and yellow longitudinal stripes. After reaching the age of six months, the pigs are no longer dependent upon their mother.

The European wild boar has been introduced into a number of places in the eastern United States, and it is so well established in the mountains of eastern Tennessee that it is now hunted there as game.

The Indian wild boar has much the same habits as the European; but since it lives in a warmer country, its hair is much more sparse and it has no underfur. Like other boars it builds shelters of grass, reeds or brush, well-roofed over, to protect itself from the elements and to rest in during the day.

The classification of the wild pigs of the East Indies and the neighboring islands is rather difficult, as natives have introduced domestic pigs about the villages and have transported them from island to island. Many of these live in a half-wild state and breed with the local wild pig.

BARBARY WILD BOAR *(Sus scrofa barbarus)*

EUROPEAN WILD BOAR. General color of the adult brown, sometimes varying to blackish or reddish. The face, cheeks and throat grizzled with whitish hairs. Bristles along dorsal line and upper neck, lengthened. A dense woolly underfur in winter. Southern and Central Europe south of the Baltic Sea.

BARBARY WILD BOAR. Slightly smaller and darker than the typical wild boar, with a shorter and less thick coat. Found in the wooded mountains of Morocco, Algeria and Tunis. Formerly another subspecies, *S. S. sennaarensis,* with a more dense and yellower coat, occurred in Egypt south to Sennar.

CHINESE WILD BOAR. There are evidently two color phases of the Chinese wild boar: one is chiefly black with a sprinkling of white hairs, and the other is more or less reddish with a mixture of black. Found throughout northern China from the coast to western Kansu and Szechwan provinces. Other subspecies are found to the north and in southern China.

INDIAN WILD BOAR. This boar differs from the European boar in having the body more sparsely covered with hair, in being without an underfur and in having a crest or mane of black bristles along the back. Varies in color from reddish brown to black intermixed with white, the old boars being the darkest. Throughout India, Burma and Ceylon. Subspecies are found in Thailand, Cambodia, Vietnam, the Malay Peninsula, and on some of the islands.

CHINESE WILD BOAR *(Sus scrofa moupinensis)*

INDIAN WILD BOAR *(Sus scrofa cristatus)*

ANDAMAN WILD PIG
(Sus scrofa andamanensis)

PYGMY HOG *(Sus salvanius)*

ANDAMAN WILD PIG. A small pig. General color black with some of the bristles of the back tipped with brownish gray. Coat long and somewhat shaggy. Found in the forests of the Andaman Islands.

PYGMY HOG. Smallest of the wild pigs. Color blackish brown due to the intermingling of the black and brown bristles. A full-grown animal stands about a foot at the shoulder. Young, striped as in the larger hogs. Found in the forests of Nepal, Sikkim and Bhutan, northern India.

BORNEAN WILD PIG, BEARDED PIG. A large pig with a high narrow body, scantily haired in the adult, and with a greatly elongated head. Cheeks have a heavy beard. About halfway between the eye and the nose there is a warty outgrowth covered with stiff bristles, very conspicuous in the male, less noticeable in the female. Found in Borneo. A close relative, *S. oi,* occurs in Sumatra.

BORNEAN WILD PIG *(Sus barbatus barbatus)*

197

BORNEAN GIANT PIG *(Sus barbatus gargantua)*

BORNEAN GIANT PIG.   Known only from a skull collected in southeastern Borneo. The largest existing wild pig. Closely related to the common Bornean wild pig, but much larger, the skull being one foot ten and one-half inches long compared with one foot seven inches, the length of the Bornean pig. Now believed to be a gigantic individual of the common Bornean pig.

JAVAN WARTY PIG.   A blackish brown pig with a slight crest of lighter brown hairs on the crown and a whitish patch on the side of each jaw. Underparts dirty white. Head long with three warts on each side of the face, the largest being just below the eye. A close relative of the Celebes pig. Occurs in Java.

PHILIPPINE WILD PIG.   A black pig with scanty hair, a few brown and white hairs among the long hairs on the forehead and upper neck, and a small patch of white hair on either jaw. A small wart between eye and nose. Closely related to the Celebes pig but skull longer and narrower Found on a number of the islands of the Philippines.

PHILIPPINE WILD PIG *(Sus philippensis)*

JAVAN WARTY PIG *(Sus verrucosus)*

198

CELEBES WILD PIG. A male specimen in the American Museum of Natural History shows this to be a small pig with sparse black hair and a whitish buff line going over the nose just below the eyes and continuing from the corner of the mouth to a white patch on the cheeks. A small wart occurs about halfway between the eye and the snout. Found in the Celebes.

CELEBES WILD PIG *(Sus celebensis)*

28
THE AFRICAN WILD PIGS, THE BABIRUSA, AND THE PECCARIES

RED RIVER HOG *(Potamochoerus porcus porcus)*

MADAGASCAR BUSH PIG *(Potamochoerus porcus larvatus)*

The bush pigs and the red river hogs are covered with long hair and many forms have been described. They vary greatly in color, many of the forms being reddish or tawny, with the old boars turning black when they reach maturity. They are more pig-like than the other African swine and generally have a tassel of hair at the end of their ears. The RUWENZORI BUSH PIG, *(P. p. intermedius)* from Uganda, is a race that is intermediate between the river hogs and the true bush pigs; and for this reason, scientists now include both groups under the one species *porcus*.

The bush pig goes about in small herds or "sounders," and often does considerable damage to native gardens. It is chiefly nocturnal. Anxious to obtain an old boar for the collection of the American Museum of Natural History, the author was advised to spend a night in a certain valley in which thousands of wild calla lilies were growing. The bush pigs are exceedingly fond of the bulbs of this lily, which is locally known as "Pig Lily." Three nights were spent among the lilies and mosquitoes before a suitable specimen was procured. The young bush pigs are brown with yellow or buff longitudinal stripes.

As the name implies, the forest hog inhabits the dense forests of Central Africa. It was first discovered in the Kakumega Forest and in the forests of Mount Kenya, Kenya. A few years after this discovery, it was found to be inhabiting the forests of the Congo and Cameroun near the Atlantic Coast.

The wart hog, one of the most grotesque of the mammals, is still common on the plains of East Africa. A number of forms are recognized which differ chiefly in size and skull characteristics.

The wart hog goes about singly or in small parties. The number of young is generally four, but two or three often occur. Like the young of the forest hog they are not striped. When trotting across the plains, the wart hog carries the tail upright with the small tuft generally pointing forward. It lives in burrows in the ground and always enters backwards so that it will be facing the entrance.

The babirusa prefers dense cover chiefly near water. It is generally found in small herds and is nocturnal in habits. Either one or two young are born at a time and they are colored like their parents. The tusks of the female are small in comparison with those of the male.

The peccaries differ in numerous ways from the true pigs, especially in tooth and bone structure. One distinction is that the upper tusk is directed downward instead of curving upward. There is a musk gland on the back about eight inches above the tail. Peccaries travel about in droves. They are found in many types of environment from heavy forest to the dry, arid country of cacti and thornbush. The young are generally two in number, although occasionally one, and are reddish in color with a black dorsal line.

SOUTH AFRICAN BUSH PIG
(*Potamochoerus porcus koiropotamus*)

RED RIVER HOG. The most handsome member of the pig family. Bright orange-red with black and white markings on the face, long white tufts on the ends of its ears, a long mane of white hairs along its back, and a long tail. The typical race is from Guinea but very close relatives are found in Liberia, Cameroun, and the Congo.

MADAGASCAR BUSH PIG. Very similar to the bush pig of the mainland but slightly smaller. A young male in the American Museum of Natural History is reddish brown overlaid with black; the shoulders, thighs and legs are black. There is a great deal of white about the head; and the mane on the neck and shoulders is a mixture of black, brown and white. There is a long pencil of white and black hair on the ear tips. Confined to the island of Malagasy.

SOUTH AFRICAN BUSH PIG. Like other bush pigs the body is covered with coarse hair, blackish brown in most of the old males; the females generally much browner, intermixed with gray. White hairs about nose. In the male a pair of warts between the eye and the nose. Found chiefly in south-eastern Africa.

FOREST HOG. A large black pig with a heavy wart below the eye. Hair long but rather scanty, the skin showing through. Tusks on the upper jaw turn upward and inward. Body heavy, legs short, tail long with a tuft. In many factors it resembles the wart hog, which is undoubtedly its nearest relative. Found in the forests from the mountains of British East Africa west through the Congo to Cameroun.

FOREST HOG (*Hylochoerus meinertzhageni*)

WART HOG (*Phacochoerus aethiopicus*)

WART HOG. General color blackish gray as the sides of the body are but scantily haired and the skin is exposed. Along the center of the back from the neck almost to the tail is a mane of long hair which hangs down on each side of the body. There are two pairs of large warts on the face, and the tusks of the upper jaw are long and turn upward and inward. Found on the plains and in the arid regions of East Africa from South Africa north to Ethiopia and the Sudan, and west to Senegal.

BABIRUSA. Body almost devoid of hair in the Celebes form, *B. b. celebensis,* the creased skin being slate-gray. The Buru form differs in having the body sparsely covered with grayish hair. The upper tusks of the male penetrate through the skin on the upper side of the snout, about halfway to the eyes, and make a backward curve sometimes coming in contact with the skin on the forehead. Found on the islands of Celebes and Buru.

BABIRUSA (*Babirussa babyrussa*)

204

COLLARED PECCARY (*Tagassu tajacu*)

COLLARED PECCARY. General color tawny and black giving a speckled appearance. Legs, feet and chest black. A light-colored collar. Found from the southwestern United States south to Patagonia.

WHITE-LIPPED PECCARY. Larger than the collared peccary. General color of upper parts black intermixed with tawny, but less so than in the collared peccary. Top and sides of the nose, chin and line along cheeks, white or yellowish white. Found from southern Mexico to Paraguay.

WHITE-LIPPED PECCARY (*Tagassu pecari*)

BACTRIAN CAMEL *(Camelus bactrianus)*

29
THE CAMELS
AND THEIR RELATIVES

The camel is used as a beast of burden throughout the desert regions of Central Asia. The hair is used for clothing and for felt coverings of dwellings; the hides, for leather; the milk and flesh, for food; and the droppings, for fuel. Before the advent of motorcars and planes, the only means of transportation was the camel caravans; and they are still extensively used. There is a possibility that the true wild camel is still found in Chinese Turkestan and southwestern Mongolia. It is a well-known fact that wild camels are found there; but whether they are feral animals, domesticated camels reverted back to the wild, or true descendants of the wild camels that once roamed the country, is a question.

The Arabian camel, or one-humped camel, is, as the name implies, characterized by a single hump. It is a much shorter-haired animal than the Bactrian camel, for it is a creature of the hot sandy wastes. The name dromedary, by which this animal is often known, refers to a certain type, the long-legged, more lightly built, riding camel; while the heavier, stockier pack animal, although with one hump, is not called by that name. The food of the camel is chiefly the leaves and twigs of bushes and trees.

The guanaco is the wild ancestor of the llama and alpaca. It is an animal of the high Andes, being often found at altitudes from 14,000 to 18,000 feet above sea level; while in the southern parts of its range it may be found at sea level. It is generally found in small herds; although in former years, herds of two or three hundred were reported.

The Incas had domesticated the llama long before the advent of the white man into South America. It was used chiefly as a beast of burden, but it also furnished a food supply and clothing. Before the importation of machinery, it was used in carrying supplies to the numerous rich mines and in bringing out the ore. To the people of the Andes Mountains it was as indispensable as the reindeer is to the Laplanders and the camel to the desert tribes. A llama can carry a load of one hundred pounds over treacherous paths at high altitudes where horses or mules are useless, for the llama is not subject to mountain sickness. A train of llamas can travel at the rate of fifteen or twenty miles a day. As a rule only the males are used as pack animals, the females being kept on pasturage and used for breeding purposes and for the fleece.

The alpaca is a domesticated form of the guanaco, which is bred for its long wool. Flocks of these animals are kept by the natives in the high mountains of Bolivia and Peru and periodically are brought down to the village for shearing. The hair of the alpaca hangs down over the sides of the animal, often reaching to the ground and concealing its legs and feet. A good alpaca may produce seven pounds of wool at a shearing. Its lofty pastures range from 12,000 to 16,000 feet. Attempts have been made to acclimate alpaca to low altitudes in other parts of the world, namely England and Australia; but it has been found that the animals do not do well at lower altitudes and these experiments have failed.

The vicuna is a much rarer animal than the guanaco and has been much persecuted for its valuable hair. Its habits are much like those of the guanaco. It goes about in small herds, often a party of females with a single male. The vicuna in-

208

habits the high plateau regions of the Andes between 12,000 and 18,000 feet. The majority of the animals are found in Peru, but some also occur in neighboring Ecuador and Bolivia. Unlike the guanaco, it does not descend to the lower altitudes. For centuries it has been hunted for its valuable fleece which is unsurpassed in softness and beauty. The fibers of the wool are strong and yet the finest of any fleece used, the diameter of a single fiber being less than half that of the fiber of the finest sheep wool. With the early Incas the fleece was reserved for the use of the rulers. Unfortunately, since the vicuna does not takes so kindly to domestication as did the guanaco, the animal had to be killed to obtain this fleece; and it became so rare that there was danger of its extinction. Fortunately the Peruvian government realized the necessity for its protection, and very stringent laws are now in effect.

ARABIAN CAMEL *(Camelus dromedarius)*

GUANACO *(Lama glama huanacus)*

BACTRIAN CAMEL. Generally of a dark brown color. This form has two humps. The Bactrian Camel existed in the wild state in Chinese Turkestan and Mongolia until recently, and it is believed that some may still be found in remote areas. Under domestication it is confined chiefly to Asia.

ARABIAN CAMEL. Of many different colors: sandy-buff, brown, white and gray being the most common. Occasionally one is black. Has but one hump. Not known in the wild state. Although not positively known, Arabia is supposed to be the ancestral home. Under domestication it has been introduced into many parts of the world, including North Africa, Australia and southern Europe and Asia.

GUANACO. General color brown with black on the face, and white underparts. Found from southern Peru, Bolivia and Patagonia south to Tierra del Fuego.

LLAMA. Occurs in a variety of colors, generally brown, often with white markings. A domesticated form of the guanaco. Found in the Andes Mountains of Peru, Bolivia, and in parts of Ecuador, Argentina and Chile.

210

LLAMA *(Lama glama glama)*

ALPACA *(Lama glama pacas)*

ALPACA. Smaller than the llama and bred for the sake of its long wool. The usual color is brown or black but white and pied-colored animals occur. A domesticated form of the guanaco. Found on the high plateaus of Peru and Bolivia.

VICUNA. Smaller and more slender than the guanaco with a shorter head and paler color, lacking the black about the face. A tuft of long white hair extends down from the chest and lower shoulders. This animal lacks the callosities which occur on the limbs of the camels and guanaco. Found in the Andes from southern Ecuador, Peru and Bolivia. It was formerly recorded from northern Chile but is no longer found there.

VICUNA *(Vicugna vicugna)*

LARGE MALAYAN CHEVROTAIN
(Tragulus napu)

30
THE CHEVROTAINS

SMALL MALAYAN CHEVROTAIN *(Tragulus javanicus)*

The chevrotain or mouse deer is not closely related to the deer but shows affiliation to the pigs and also to the camels, although it lacks the upper incisors and chews its cud. It prefers thick wooded or bushy country but at night is sometimes seen by the lights of a car, along the edges of unfrequented roads. As it scurries off through the grass or underbrush, it reminds one of a rabbit. The chevrotain of India and Ceylon is distinctive in the spotting of the back and sides; while the forms found farther to the east are more uniform in coloring. Over fifty subspecies of the large and the small Malayan chevrotains are recognized, differing chiefly in the intensity of color and also in the size and markings. Their extensive range includes many of the smaller islands; and some of these islands have a subspecies characteristic of that island alone. The smallest form of the small Malayan chevrotain rivals the royal antelope of West Africa for the distinction of being the smallest of the ungulates. There is one species, the PHILIPPINE CHEVROTAIN, *T. nigricans,* found on Balabac Island in the Philippines.

The large Malayan chevrotain differs from its smaller relative in having proportionately longer hoofs, and is said to be more partial to swampy country.

The males of all chevrotains have long upper canine tusks which protrude slightly below the upper lip. Through the greater part of the year the animal lives a solitary life but it sometimes is found in pairs, and the mother may be found escorting one or two young.

This animal is so similar in actions, shape and size to the South American agouti, a large rodent related to the guinea pig, that when I surprised one crossing a woodland road in Vietnam, my mind immediately returned to a similar setting of a few years before in British Guiana when I observed my first agouti. The chevrotain walks on the tips of its hoofs in a peculiar stiff-legged manner, giving the impression that its feet are very tender. Like some of the African duikers it will sometimes climb into the branches of a fallen or vine-covered tree and even seek safety in a hollow some distance above the ground. If taken young, the chevrotain becomes very tame and makes an engaging pet. It is said to have bred in captivity.

The water chevrotain is a larger animal than the mouse deer and very different in habits. Three forms are recognized: the typical form of Gambia and Sierra Leone; BATES' WATER CHEVROTAIN, *H. a. batesi,* found in Cameroun; and the CONGO WATER CHEVROTAIN, *H. a. cottoni.* Bates' water chevrotain has a yellow wash in the spots and stripes of the body. In the Congo form, the spotting is much less pronounced. This animal is found in the Ituri forest of the northeastern Congo.

As the name implies, the water chevrotain is partial to water and is generally found along the banks of rivers and streams in the forests. It seeks protection by plunging into the water and hiding in the aquatic vegetation. It is chiefly nocturnal and rather shy so that it is seldom observed. The natives consider the flesh a great delicacy and catch the animal either in snares or by driving it into nets.

The male is generally darker in color than the female and is further distinguished by the upper canine teeth which form tusks about an inch long and project below the upper lip.

214

Although it has been stated that two young are born at a time, the large series in the American Museum of Natural History seems to prove that only a single young is the rule. The stomach contents showed that the animals had been feeding on grass, aquatic vegetation, leaves from bushes, and fruits.

The water chevrotain resembles the pigs in having the second and fifth metacarpal bones well developed; and the second and third, instead of being fused together to form the cannon bone, as is the case with most of the ungulates, are separated. The metatarsal bone of the hindleg is fused together but not to the same degree as in the other ungulates.

LARGE MALAYAN CHEVROTAIN. Upperparts brown or yellowish brown, grayer on the sides. Underparts white, generally with a brown median line on chest and lower abdomen. Throat and foreneck generally with five white bands. Tail brown above, white below. Found in the Malay Peninsula and southern Tenasserim south through Sumatra, Java, Borneo and neighboring islands. *T. stanley anus* is slightly brighter in color and is restricted to the Malay Peninsula and adjacent islands.

SMALL MALAYAN CHEVROTAIN. Above brown with rufous tinge. Sides paler. Nape and upper parts of neck dark. Underparts white, sometimes mixed with light rufous, with a narrow median line of brown. Three white stripes extending up the throat and joining in a large white patch on upper throat and chin. Tail reddish brown above, white below. Found from Indochina and Siam south throughout the Malay Peninsula and Tenasserim, Sumatra, Borneo, Java and many of the neighboring islands.

INDIAN CHEVROTAIN
(Tragulus meminna)

INDIAN CHEVROTAIN. Olive-brown speckled with yellow. Flanks have longitudinal rows of white or buffy spots, which form distinct bands. Three white stripes on the throat. Underparts white. Found in the forested area of Ceylon and southern India.

WATER CHEVROTAIN. A reddish-brown animal with under surface of chin, throat and chest white. Three distinct white bands running from the shoulder along the flanks; spots on back forming distinct white bands; tail brown above. Found in Gambia and south into Liberia and Cameroun and east into the Ituri Forest of the Congo.

WATER CHEVROTAIN *(Hyemoschus aquaticus aquaticus)*

31
THE ONE-HORNED
RHINOCEROSES

INDIAN RHINOCEROS *(Rhinoceros unicornis)*

The Indian rhinoceros is the largest of the rhinoceroses found in Asia. Its strange form, ungainly head, thick skin with the numerous folds and the horn on the nose, make one think of some prehistoric creature. Indeed it is a creature from the past, for fossil remains found from the Pliocene and Pleistocene eras show that at one time the rhinoceroses were much more common than they are today and that there were many more kinds. They were found in Europe, Asia, Africa and North America. Formerly the Indian rhinoceros was found throughout most of India; but now its range is limited to northern Assam, northern Bengal and eastern Natal. In a recent survey it was estimated that there were fewer than three hundred of these animals in existence. The great majority of these rhinoceroses are found in Assam, where, fortunately, most of them are on reserves: the Kaziranga Sanctuary in Sibsagar District, Assam, is the most important. Since it is doubtful if there are fifty living rhinoceroses left in either Nepal or Bengal, it is unlikely that the Indian rhinoceros can long survive without the strictest kind of protection.

All of the Asiatic rhinoceroses have suffered much the same fate and all are on the verge of extinction. Certain parts of this animal, including the blood, bones, skin, but chiefly the horn, have for years been believed by the natives and especially by the Chinese to contain medicinal properties. In former years, even in England, it was believed that poisoned wine poured into a cup made of rhinoceros horn could easily be detected as it would foam immediately. Another version was that the power of such a cup would make the poison harmless. Needless to say these cups were in great demand.

On account of their beliefs, the Chinese are willing to pay huge sums for a dead rhinoceros, or for its horn; and in consequence this animal has been diligently hunted and is still poached on the reserves where it is making its last stand.

The home of the Indian rhinoceros is the tall grass and canes which grow to a height of over ten feet on the plains of its habitat. Seldom does it leave its fastnesses, for it has learned that they afford it vital protection. Nor is it ever found far from water, for it enjoys wallowing in the mud.

The two lower incisor teeth of the Asiatic rhinoceroses are sharply pointed tusks which are too small to be visible when the animal's mouth is closed; but it is with these tusks that the animal does its fighting and not with its horn as many people suppose. The African rhinoceroses do not have such tusks, and their horns are their weapons. The horn of the rhinoceros is made of solidified hair-like fibers which grow from the skin. When the animal is skinned, the horn can be removed from the skull with the skin since it is not solidly attached.

The Javan rhinoceros looks very much like a small Indian rhinoceros, but the texture and the fold of the skin on the foreshoulder serve to distinguish it. In the Indian rhinoceros this fold has a backward sweep as it reaches the upper shoulder, but it does not quite join the fold that continues over the back just behind the shoulder. This backward sweep of the forward fold gives the shoulder "plate" the appearance of a triangle.

In the Javan rhinoceros the front fold extends up over the foreshoulder continuously to the opposite side and is more or less parallel to the fold of the rear shoulder. The female differs from the female Indian rhinoceros in that it has no horn.

The Javan rhinoceros is at the present time one of the rarest of animals. From its former extensive range it is now restricted to a few isolated spots. The very few still left in Burma are restricted to the Kahilu Game Sanctuary in Lower Burma, and in 1937 it was believed that about four animals still remained there. It is doubtful if this rhinoceros is still found in the Malay States, although it was thought in 1937 that possibly a half dozen might have survived there. In Sumatra a few of these rhinos still exist but in very small numbers, and they are steadily decreasing; while in Java, although protected since 1909, they are illegally killed. There are a few in the Bantam District on the western tip of the island, but the political disturbances prevent their being afforded the necessary protection.

The Javan rhinoceros prefers thick jungle and is even found in the mountainous forests. Like the Indian rhinoceros it spends much of its time wallowing in the mud and water and often these characteristic mudholes reveal its presence.

JAVAN RHINOCEROS *(Rhinoceros sondaicus)*

INDIAN RHINOCEROS. Skin dark gray with a fringe of black hair on ears and end of tail. Differs from the Javan rhinoceros in its larger size and the fold of the skin on the foreshoulders, which extends backward toward the rear of the shoulder. The thick hide is covered with numerous rounded tubercles which look like rivet heads. Both sexes carry horns, the females generally smaller. Now restricted to limited areas in Assam, Bengal, and eastern Nepal.

THE JAVAN RHINOCEROS. Similar in color to the Indian rhinoceros but smaller; and the skin lacks the rounded tubercles; the finer cracks of the skin form a mosaic pattern. Fold at the front of the shoulder extending up over the back, parallel to the fold behind the shoulder. Formerly found from eastern India south through the Malay Peninsula to Sumatra and Java. Now restricted to a few isolated regions in Burma, Malay Peninsula, Sumatra and Java.

222

32
THE TWO-HORNED RHINOCEROSES

SUMATRAN RHINOCEROS (*Didermocerus sumatrensis sumatrensis*)

The Sumatran rhinoceros has shared almost the same fate as the other Asiatic rhinoceroses. The incessant hunting of the animal for its horns has wiped it out from most of its original range. A few are still said to exist in Sumatra, but judging from reports, they are very scarce. In Borneo it still occurs in the mountain forests, but it is still killed by the natives and is very rare. Undoubtedly, more are found in the mountainous parts of the Malay Peninsula than anywhere else in the world. There are infrequent reports of it from Thailand; a few still inhabit Burma; but it is now considered doubtful if any remain in Assam.

The Sumatran rhinoceros lives in dense mountain forests and in former years wore deep trails along the mountainsides. The thick cover makes it a difficult animal to hunt, but the natives dig pits in the trails and in this way capture the animal. Like the other Asiatic rhinoceroses it makes deep wallows in the mud where the natives wait for its coming.

The young Sumatran rhinoceros is covered with curly brown hair, but this disappears as the animal grows older. Two subspecies of this animal are recognized: the hairy-eared form from Assam and northern Burma, which now may be extinct, and the MALACCAN RHINOCEROS, *D. s. niger,* of southern Burma and the Malay, which is darker with the bristle-like hairs of the body black.

The black rhinoceros of Africa is the most common rhinoceros today. Although it has been exterminated in much of

South Africa, it is still found in goodly numbers in Tanzania and Kenya. This animal seems to prefer the dry open country, but it also enters the mountainous forests to some extent. It never is found far from water, for it delights in mud baths. This animal is a browser, feeding on the leaves and twigs of certain trees: it seems to prefer the thorny acacias. The pointed upper lip is of great assistance in pulling food into the mouth. Like all the rhinoceroses its eyesight is poor but its sense of smell and hearing is acute. It is the most pugnacious of the rhinos and will sometimes charge without provocation.

Once when we were tracking elephants with two natives, our path led us past a large acacia tree under which three lions were resting. We kept on our way undaunted until we approached close to the lions. The natives shouted and waved their arms and the lions bounded away. About a mile farther on, we came across a single rhinoceros enjoying his siesta. This time the leading native deviated from the trail, giving the rhino a wide berth, and only after we were safely past did we pick up the elephant trail again. This shows how great a respect the natives have for this animal.

The white rhinoceros was first discovered and described in South Africa. Formerly it was common over much of the open country; but its numbers were so greatly reduced that at the present time all of the remaining white rhinoceroses are found in the Umfolozi and Hluhuwe Reserves in Natal. There are estimated to be about two hundred survivors. In 1900 Major A. St. H. Gibbons sent to Europe the skull of a white rhinoceros which had come from the west bank of the Nile in the southern Sudan. Thus, at the time when the white rhinoceros was fast disappearing, a new race was discovered two thousand miles to the north.

Unlike the black rhinoceros, the white rhinoceros is a grazer, preferring the open grassy country. It is more social than the black rhinoceros, small parties associating together. It is also of a much more peaceful disposition. In walking, it holds its heavy head very low, the nose almost touching the ground.

HAIRY-EARED RHINOCEROS *(Didermocerus sumatrensis lasiotis)*

SUMATRAN RHINOCEROS, ASIATIC TWO-HORNED RHINOCEROS. Smallest of the rhinoceroses. General color brownish gray, thinly covered with short black hair. Ears edged with short black hair. As the name implies, it has two horns; and like the other Asiatic rhinoceroses, the lower incisors form two small tusks. Formerly found from Bengal and Assam, through Burma, Thailand, the Malay Peninsula to Sumatra and Borneo. The typical form is restricted to the two islands.

HAIRY-EARED RHINOCEROS. Subspecies of the Sumatran rhinoceros, characterized by the long hair which fringes the ears and the lower portion of the tail. Skin grayish, covered with inch-long dark brown bristles. Skin with two folds extending over the body but not so pronounced as in the one-horned rhinoceros. The northern form of the Sumatran rhinoceros, found in Bengal, Assam and northern Burma.

227

BLACK RHINOCEROS *(Diceros bicornis)*

BLACK RHINOCEROS. Dark brown in color with a relatively small head and two horns. The muzzle is rather narrow and the upper lip is triangular with a hooked tip. Formerly found throughout most of East Africa from Ethiopia to the Cape, and southwestern Angola. Now exterminated in the southeastern part of its range. In the northwest it extends to Lake Chad.

WHITE RHINOCEROS, SQUARE-MOUTH RHINOCEROS. General color slightly lighter and grayer than the black rhinoceros. Head elongated, the two horns long and the lips broad. Formerly found in South Africa north to the Zambesi. Now restricted to two reserves in Natal. The northern subspecies, *C. s. cottoni*, is found from the west bank of the upper Nile in the southern Sudan, to the border of Central African Republic and the northeastern Congo and northern Uganda.

WHITE RHINOCEROS (*Ceratotherium simus simus*)

33
THE TAPIRS

MALAY TAPIR *(Tapirus indicus)*

The tapir has the distinction of being one of the oldest forms of mammals which has remained unchanged. Fossil remains of the tapir found in the Pliocene Age differ so slightly from the tapirs of today that it is placed in the same genus. In fact, an ancestral tapir found in the Miocene Age was so similar that for many years it, too, was placed in the genus, *Tapirus*. In former years tapirs were much more common than today and they had a much more extensive range. Tapirs of the genus *Tapirus* inhabited Europe and Asia during the Pliocene Age and found their way into North and South America during the Pleistocene Age, thus accounting for the way the tapirs of the Old and new World kept their close relationship although living so far apart.

The Malay tapir, sometimes called the blanket tapir, on account of the white marking on the back which gives the appearance of a black animal covered with a white blanket, is the only tapir found in the Eastern Hemisphere, all other modern forms being found in Central and South America. Like all tapirs, the Malay tapir is a shy animal and, being nocturnal and living in dense swamps and forests, it is seldom observed. It is generally a solitary animal, though occasionally two adults may be found together or a mother with her spotted young may be seen. Although shy, if captured young it becomes very tame and does very well in captivity. On account of this fact and their interesting appearance there is a great demand for tapirs in zoological gardens. While at Singapore, I was interested in observing a pair of these animals in a private zoo. Each morning they were given their full liberty and would stroll down to the river, where they would bathe; but in about an hour's time they would return to their pen on their own initiative.

The South American tapir is the smallest of the tapirs. Like most tapirs, it prefers to live along the banks of rivers or lakes, but it is also known to ascend the mountains to a considerable altitude. It feeds on water plants as well as upon leaves and other herbage and fruits.

The natives generally hunt the tapir with the aid of dogs. The disturbed animal seeks the nearest water for protection. Here the natives wait for its coming, and if the water is not too deep, they are able to dispatch it; but if the water is of sufficient depth, the animal frequently can make its escape. The great weight of the tapir enables it to push its way through the thick underbrush at considerable speed, much faster than it can be followed.

As the tapir is the largest animal to be found over much of its territory, it has few enemies. Besides man it is preyed upon by the jaguar which seems to hold its flesh in high esteeem.

A single spotted young is born in a secluded place in the forest and is almost immediately able to accompany its mother.

The mountain tapir is an inhabitant of the high forests of the mountains of Colombia and Ecuador. It appears to be more common in the latter country and has been recorded at an altitude of over 14,000 feet. It forms paths through the thick cover on the mountainside and is never found very far from water, often frequenting the mountain lakes.

Like all tapirs, Baird's tapir lives in the forested regions, always near water. It is found from sea level up to over five thousand feet in the mountain forests. On certain mountainsides its trails, which generally lead to water, are very much in evidence and are often over rough terrain, which speaks well for the animal's climbing ability.

Baird's tapir is the form found in the Canal Zone of Panama and is sometimes seen along the shores of Gatun Lake. Occasionally it visits the Scientific Research Island of Barro Colorado.

Dow's tapir, at one time thought to be a distinct species found in Honduras and Guatemala, is now believed to be the same as the Baird's tapir.

SOUTH AMERICAN TAPIR *(Tapirus terrestris)*

MALAY TAPIR. Head, neck, shoulders, fore- and hind-legs and lower thighs brownish black. Rest of body grayish white. Young, dark brown with numerous yellow and white streaks and spots. Found from southern Lower Burma and southern Thailand south through the Malay Peninsula and Sumatra.

SOUTH AMERICAN TAPIR. The smallest of the tapirs, of a uniform dark-brown color, with a short upright mane on the neck and crown of the head. As with the other tapirs the toes of the front feet number four and the toes of the hind feet number three. Found from northern South America as far south as southern Brazil, Paraguay and northern Argentina.

MOUNTAIN TAPIR. Differs from the other tapirs in being covered with long kinkly hair of a black color, the individual hairs being about an inch long. Like the other tapirs, the outer edge of the ears is white and there is a white area at the mouth which includes both lips. It lacks the crest and mane of the South American tapir. Found in the mountains of central Colombia and Ecuador and posisbly to northern Peru.

MOUNTAIN TAPIR (*Tapirus roulini*)

BAIRD'S OR GIANT TAPIR. About the size of the Malay tapir. General color blackish brown, lighter on head and cheeks. The edges of the lips and tips of the ears, patch on the throat and chest, white or whitish. Like the other tapirs, the young are marked with white stripes and spots. Found in Central America from Mexico south through Panama.

IRD'S TAPIR *(Tapirella bairdii)*

PRZEWALSKI'S WILD HORSE *(Equus przewalskii)*

34
THE WILD HORSE
AND THE ASSES

Przewalski's wild horse is of special interest as it is the only true wild horse in existence today. The so-called wild horses of our western plains are feral animals, horses which have been in domestication themselves or whose ancestors have been under domestication and have broken away and reverted to the wild.

Przewalski's wild horse looks like a small stockily-built horse with a large head, thick neck, and upstanding mane, but lacking a forelock. Originally this horse was common on the plains of Mongolia in the Gobi Desert, and in Chinese Turkestan; but in more recent years it has been restricted to the country in the vicinity of the Altai Mountains, especially in Dzungaria. At the present time the herds are greatly reduced in numbers. Unfortunately, feral horses roam over much of this country, and as the herds of wild horses and the domestic horses often run together and they readily interbreed, pure Przewalski's wild horses are becoming scarce and the world is losing a most interesting animal.

The onager of Iran and Afghanistan is closely related to the INDIAN WILD ASS, *E. h. khur* of the deserts of Northwest India, the Rann of Cutch and Baluchistan. This animal, however, lacks the shoulder stripe which is generally found on the Iran form. Another close relative was the SYRIAN WILD ASS, *E. h. hemippus,* which at one time was found in Syria, northern Arabia and eastern Iran. It was the smallest of the wild asses and was of a gray color. It is now believed to be extinct.

The Mongolian wild ass has been greatly reduced in numbers, and is now restricted to the country east of the Altai Mountains; it is common only in the Gobi Desert. Here Dr. Roy Chapman Andrews found it in numbers: he writes that they were abundant especially about Tsagan Nor and at his camp at Loh. This camp was situated on a flat gravel plain near a spring where the animals came to drink; and this undoubtedly accounted for their numbers. Farther south at Tsagan Nor, they were even more numerous, and one herd of over one thousand individuals was observed. In the late spring, just before the foals are born, these large herds break up into smaller parties. This animal is fleet of foot and possesses great stamina. Dr. Andrews tested its speed and endurance from an automobile, and he stated that when it was forced to, this animal could travel at the rate of forty miles an hour for a mile or so, but that it could not long stand the pace and would slow down to thirty-five miles per hour. One male was eventually run to a standstill after a chase of twenty-nine miles, during which he averaged thirty miles per hour for sixteen miles.

The kiang is the largest of the wild asses, with large hoofs. Its home is on the bleak plateaus of Tibet. For protection the coat of this animal has very long hair in winter. It is a close relative of the Mongolian wild ass with much the same habits, but is brighter in color. It is the one wild ass that still is found in large numbers.

The African wild asses are grayer in color than the Asiatic forms and have much longer ears. The Nubian wild ass is considered to be the ancestor of the domestic donkey. Like all the wild asses, they prefer to live in the most arid regions, but they keep themselves sleek and fat on the scanty vegetation. Unlike most of the domestic donkeys, the wild asses are constantly on the alert, fleet of foot, and very difficult to approach. Unfortunately the wild ass is a very rare animal in Africa and is rapidly approaching extinction.

ONAGER *(Equus hemionus onager)*

PRZEWALSKI'S OR MONGOLIAN WILD HORSE, TARPAN. Upperparts in summer reddish brown becoming lighter on the underparts with white about muzzle. The erect mane is dark brown and there is no forelock as in the domestic horse. A dorsal stripe and a shoulder stripe present. Winter coat long and lighter in color, and dorsal and shoulder stripes lacking or very indistinct. Lower parts of legs black. Tail long, hair shorter at base but hair of the rest of tail long and black. Found in Dzungaria and the plains regions in the vicinity of the Altai Mountains, western Mongolia.

ONAGER, PERSIAN WILD ASS. Smaller and paler in color than the Mongolian wild ass. General color light sandy-red with a light brown dorsal stripe. On each side of the dorsal stripe there is a whitish stripe which meets the light color of the hind quarters. There is generally a shoulder stripe. Underparts including lower throat, chest, legs, lower belly and shoulder and buttocks, white. In winter the hair is longer and grayer with the white area well defined. Found in northern Iran and Afghanistan. Close relatives are found in neighboring countries.

MONGOLIAN WILD ASS *(Equus hemionus hemionus)*

MONGOLIAN WILD ASS, KULAN, CHIGETAI. Slightly smaller and lighter in color than the kiang. In summer the general color is reddish yellow, grayer in winter. Nose, inner parts of forelegs and hind legs, lower hind quarters, white. Legs and underparts lighter than body. A narrow black dorsal stripe from withers to base of tail. Formerly found from southern Siberia and western Manchuria west across Mongolia to Chinese Turkestan. Now restricted to the more inaccessible regions of Mongolia. Still numerous in the Gobi Desert.

KIANG, TIBETAN WILD ASS. In summer the upperparts are reddish while the legs, underparts, buttocks and nose are grayish white, in sharp contrast to the upper body. In winter the hair is long and the contrast is not so great. Largest of the wild asses. Found on the high plateaus of Tibet.

240

KIANG *(Equus hemionus kiang)*

ABYSSINIAN WILD ASS (*Equus asinus somaliensis*)

ABYSSINIAN OF SOMALI WILD ASS. A slightly heavier animal than the Nubian wild ass. General color reddish gray, nose gray, a dark dorsal stripe which is not very pronounced or sometimes lacking, dark bands on the legs. Generally no shoulder stripe. In summer coat is grayer. Found in Somalia and eastern Ethiopia.

NUBIAN WILD ASS. General color grayish fawn; muzzle, ring around eye, and underparts, white. A narrow dorsal stripe and a distinct shoulder stripe. Legs with no striping. A short upright mane, dark brown. Found in Sennar and Nubia.

NUBIAN WILD ASS (*Equus asinus africanus*)

241

35
THE ZEBRAS

QUAGGA (*Equus quagga*)

The range of the quagga was limited to the southeastern part of South Africa as far north as the Vaal River in Orange Free State. The settlement of this country spelled the doom of this animal; and it became extinct in 1875 when the sole survivor died in the Berlin Zoological Garden. The quagga formerly occurred in large numbers, generally going about in small herds of fifteen or twenty individuals and often associating with the white-tailed gnu. The Boer farmers killed many of them to supply meat to their native workers, and later the demand for their hides led to their extinction. At the present time there are only about twenty known skins and a few skulls in existence, principally in the museums of Europe. None are to be found in any collection in America.

The Burchell's zebras, sometimes known as the bontequaggas, as a group include all forms of the common zebra and are now considered one species. There is so much variety in the striping and color within this group that over twenty subspecies have been described. Oftentimes animals typical of the different subspecies could be found in the same herd, a fact which made classification very difficult. In 1936, Angel Cabrera wisely reduced the number to four subspecies, and his classification and key are used here. It is interesting to note that in the markings on the legs the zebras are just the opposite of the giraffe. In the former the lighter or unmarked legs are found in the southern forms and grow darker toward the north, while in the giraffes the northern forms have the lighter or unmarked legs while those farther south are marked. On the plains of East Africa zebras are one of the most common of the larger animals, frequently herding with wildebeests and gazelles. They are perhaps more commonly preyed upon by the lion than are any of the other game.

The mountain zebra prefers hilly and broken ground. Like all of the larger animals of South Africa it was greatly persecuted until at the present time it is doubtful if as many as one hundred individuals remain. The South African government has made the Cradock Mountain Zebra National Park a sanctuary for the preservation of this zebra. Hartmann's mountain zebra is more fortunate and still occurs in numbers throughout most of its range.

Compared with the common zebra, Grevy's zebra has a limited range, preferring the semi-desert region of northeastern Kenya and neighboring Ethiopia and Somalia, where the country is hot and dry and the vegetation is composed of thornbush and acacia with an occasional plain of parched grass. It generally goes about in small herds. It is more wary than the common zebra.

244

BURCHELL'S ZEBRA (*Equus burchellii burchellii*)

DAMARALAND ZEBRA *(Equus burchellii antiquorum)*

QUAGGA. General color of the body brown or bay becoming lighter on the rear parts. Underparts, tail and legs white. A broad dark dorsal line. Dark brown and white stripes on the head, neck and anterior part of the body, fading out from the shoulders to the posterior. Nose black. Formerly found in Orange Free State and northern and central parts of Cape Colony. Now extinct.

BURCHELL'S ZEBRA. Legs entirely unmarked with any striping, although in some specimens there may be a faint tracing just above the hock. Shadow stripes generally present. This type race of Burchell's zebra formerly was found in the Orange Free State and southern Botswana. Now extinct.

DAMARALAND ZEBRA. Legs have some striping but never completely marked to the hoof. Shadow stripes generally present. Found from southern Angola and South-West Africa across Botswana to Transvaal and Zululand.

SELOUS' ZEBRA. Legs completely striped to the hoofs, shadow stripes generally absent or very few. Stripes and interspaces on neck and body, narrow and numerous; neck stripes ten to thirteen; vertical body stripes four to eight. Found from Rhodesia and Mozambique north to southern Malawi.

SELOUS' ZEBRA *(Equus burchellii selousii)*

EAST AFRICAN ZEBRA *(Equus burchellii bohmi)*

EAST AFRICAN ZEBRA. Like Selous' zebra in having the legs completely striped to the hoofs and generally without shadow stripes, but the stripes of the neck and body wider and less numerous. Neck stripes seven to ten and vertical body stripes three to four. Found throughout Zambia, west of the Loangwa River and from northern Malawi north through Tanzania, Kenya, Uganda, southern Sudan, southern Ethiopia and southwest Somalia.

MOUNTAIN ZEBRA. A small stockily-built zebra with very wide stripes, the black stripes being the broader. Stripes of the face from below the eye to the nose are of a reddish-brown color with two patches at the nostrils, of the same color. The nose itself is black. A series of transverse stripes extends over the haunches to the base of the tail forming the "gridiron pattern" characteristic of this zebra. On the throat of the male there is a small dewlap. Originally found in the highlands of Cape Colony but now practically extinct.

MOUNTAIN ZEBRA *(Equus zebra zebra)*

HARTMANN'S MOUNTAIN ZEBRA *(Equus zebra hartmannae)*

HARTMANN'S MOUNTAIN ZEBRA. Slightly larger than the mountain zebra with broader white stripes on the haunches, giving the appearance of a much whiter animal. Found in the mountain ranges of western South-West Africa and north into western Angola where it is limited to the mountains near the coast.

GREVY'S ZEBRA. The largest of the zebras, very mule-like in appearance with large, broad ears and a large head. The stripes numerous and very narrow and not continuing under the body, leaving the underparts white and unstriped. The legs striped to the hoofs. Found in southern Ethiopia east to the Somalia border and south into northern Kenya.

GREVY'S ZEBRA *(Dolichohippus grevyi)*

247

There are no two individual zebras with identical markings.
The diagnostic markings of the different species and sub-species of the zebras are shown on pages 250-253. Pages 250-251 illustrate examples of the four subspecies of the Burchell's zebra group and show the great variations that may be found in each. Variations such as these may be found in a single herd, although animals inhabiting different sections of the country tend to have certain characteristics in common. The group in the center on page 251 represents variations of the true Burchell's zebra, now believed extinct. As can be noted, the entire leg lacks markings, although occasionally there may be faint traces of stripes just above the hock. The first example shows the extreme in extensive white area: it was used as the type of subspecies *paucistriatus* and is in the Mainz Museum. The second is a male that was in the Knowsley Hall Park in 1845 and shows faint striping above the hock. The third, which also shows this striping, lived in the Amsterdam Zoological Gardens. The fourth is a male which was formerly in the Dresden Zoological Gardens.

The group at the top of page 250 is the Damaraland zebra; it has some striping on the legs but this does not extend to the hoofs. The first is a female originally from the country west of Lake Ngami in northern Botswana; it later lived in the Madrid Zoological Park. The second is a male that came from Reitfontein West, northern South-West Africa, and is now in the Munich Museum. The third is a female from the eastern Transvaal. The fourth was from Zululand and is the type of *wahlbergi*.

The central group on page 250 represents Selous' zebra, with numerous stripes along the neck and body and markings extending to the hoofs. The first originally came from near Tate, Mozambique, and is the type of *foai;* it is now in the Paris

Museum. The second, the type of *annectens,* came from Fort Jameson, Zambia, and is now in the Tring Museum. The third, a female, originally came from Nyspindire Tendo, Mozambique, and is now in the Rouen Museum.

The group at the bottom of pages 250-251 is the East African zebra, with fewer stripes on the neck and sides, the stripes extending to the hoofs. The first is a male from east of Lake Naivasha, Kenya, now in the Basel Museum. The second formerly lived in the Royal Zoological Gardens of Scotland and is said to have come from southern Ethiopia. The third, the type of *zambeziensis,* from Morotsiland, Zambia, is now in the Paris Museum. The fourth, a very dark broad-striped individual showing bohmi characteristics, at one time lived in the New York Zoological Park and was believed to have come from the Kilmanjaro region. The fifth, here shown as a foal, came from east of Lake Naivasha, Kenya, and is in the Basel Museum.

Except for the Burchell's zebra in the Amsterdam Zoological Gardens and the East African zebras from southern Ethiopia and the Kilimanjaro region, the patterns are from Angel Cabrera, *Subspecies and Individual Variation in the Burchell Zebras.*

The markings of the species of zebras not included in the Burchell group are shown on pages 252-253.

DAMARALAND ZEBRA *(Equus burchellii antiquorum)*

SELOUS' ZEBRA *(Equus burchellii selousii)*

EAST AFRICAN ZEBRA *(Equus burchellii bohmi)*

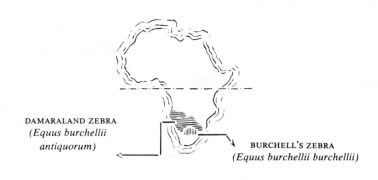

DAMARALAND ZEBRA
*(Equus burchellii
antiquorum)*

BURCHELL'S ZEBRA
(Equus burchellii burchellii)

:HELL'S ZEBRA *(Equus burchellii burchellii)*

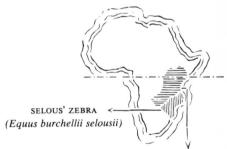

SELOUS' ZEBRA
(Equus burchellii selousii)

EAST AFRICAN ZEBRA
(Equus burchellii bohmi)

251

MOUNTAIN ZEBRA (*Equus zebra zebra*)

QUAGGA (*Equus quagga*)

252

GREVY'S ZEBRA (*Dolichohippus g*

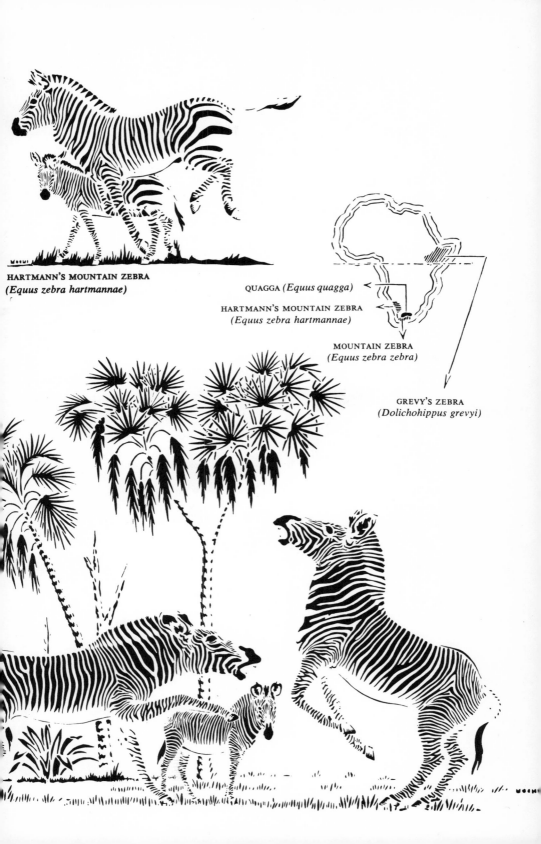

HARTMANN'S MOUNTAIN ZEBRA
(Equus zebra hartmannae)

QUAGGA (*Equus quagga*)

HARTMANN'S MOUNTAIN ZEBRA
(Equus zebra hartmannae)

MOUNTAIN ZEBRA
(Equus zebra zebra)

GREVY'S ZEBRA
(Dolichohippus grevyi)

BIBLIOGRAPHY
INDEX

BIBLIOGRAPHY

ALLEN, G. M.
>   1938. *The Mammals of China and Mongolia.* 2 vols. American Museum of Natural History, New York.

>   1939. *A Checklist of the African Mammals.* Bulletin of the Museum of Comparative Zoology, vol. 83, Cambridge, Mass.

>   1942. *Extinct and Vanishing Mammals of the Western Hemisphere.* Special Publication No. 11, American Committee for International Wild Life Protection, New York.

ANTHONY, H. E.
>   1928. *Field Book of North American Mammals.* G. P. Putnam's Sons, New York.

BOONE AND CROCKETT CLUB
>   1939. *North American Big Game.* Charles Scribnei's Sons, New York.

CABRERA, ANGEL
>   1936. *Subspecies and Individual Variation in the Burchell Zebras.* Journal of Mammalogy, vol. 17, pp. 89-112.

CHASEN, F. N.
>   1940. *A Handlist of Malaysian Mammals.* Bulletin of the Raffles Museum, No. 15, Singapore.

EDMOND-BLANC, FRANÇOIS
>   1947. *A Contribution to the Knowledge of the Cambodian Wild Ox or Kouproh.* Journal of Mammalogy, vol. 28, pp. 245-249.

ELLERMAN, J. R. and T. C. S. MORRISON-SCOTT
>   1951. *Checklist of Palaearctic and Indian Mammals, 1758-1946.* British Museum of Natural History, London.

HARPER, FRANCIS

    1945. *Extinct and Vanishing Mammals of the Old World.* Special Publication No. 12. American Committee for International Wild Life Protection, New York.

KINLOCK, ALEXANDER, A.

    1892. *Large Game Shooting in Thibet, the Himalayas, Northern and Central China.* Thocker, Spink and Co., Calcutta.

LYDEKKER, R.

    1898. *Deer of All Lands.* Rowland Ward, Ltd., London.

    1913-1916. Catalogue of Ungulate Mammals in the British Museum, 5 vols. British Museum of Natural History, London.

MILLIAS, J. G.

    1895. *The Breath of the Veldt.* Henry Sotheran and Co., London.

ROOSEVELT, THEODORE and EDMUND HELLER

    1914. *Life Histories of African Game Animals.* 2 volumes. Charles Scribner's Sons, New York.

SCLATER, P. L. and O. THOMAS

    1894-1900. *The Book of the Antelopes,* 4 volumes. R. H. Porter, London.

SETON, E. T.

    1928. *Lives of Game Animals.* 8 volumes. Doubleday, Doran and Co., New York.

SHORTRIDGE, G. C.

    1934. *The Mammals of South West Africa.* 2 volumes. William Heinemann, Lts., London.

SIMPSON, G. G.

    1945. *The Principles of Classification and a Classification of Mammals.* Bulletin of the American Museum of Natural History, vol. 85, New York.

STROOCK, S. I.

    1937. *Llamas and Llamaland.* S. Stroock and Co., Inc., New York.

    1937. *Vicuna. The World's Finest Fabric.* S. Stroock and Co., Inc., New York.

WARD, ROLAND

    1928. *Records of Big Game.* Roland Ward, Ltd., London.

INDEX OF ANIMALS PICTURED

(Page numbers in italic type indicate position of picture)